NATIONAL DEFENSE RESEARCH INSTITU

T0108992

Improving Decision Support for Infectious Disease Prevention and Control

Aligning Models and Other Tools with Policymakers' Needs

David Manheim, Margaret Chamberlin, Osonde A. Osoba, Raffaele Vardavas, Melinda Moore

Prepared for the Office of the Secretary of Defense

For more information on this publication, visit www.rand.org/t/RR1576

Library of Congress Cataloging-in-Publication Data
ISBN: 978-0-8330-9550-3

Published by the RAND Corporation, Santa Monica, Calif.
© Copyright 2016 RAND Corporation
RAND® is a registered trademark.

Cover: Scientist, © iStock: Reptile8488; Bacteria, © Fotolia: lotus_studio.

Support RAND
Make a tax-deductible charitable contribution at
www.rand.org/giving/contribute

www.rand.org

Preface

In 2015, the U.S. Department of Defense participated in a federal interagency disease-forecasting challenge to help identify suitable forecasting models for dengue fever, a debilitating mosquito-borne infectious disease that is common across the tropics and subtropics and thus highly relevant to military medical readiness. As a result of that disease-forecasting challenge, Department of Defense leaders recognized the need to better connect modeling to the policy context, and that need gave rise to the present report.

Thus, this report describes models and other decision-support tools that can help provide answers to real-world questions about infectious disease prevention and response. The intended audience includes technical experts and the policymakers whom those experts can support. This overview should help modelers with the critical task of understanding both the questions that policymakers want answered and when those answers are needed. Similarly, as we explain, policymakers can benefit from a basic understanding of the capabilities and limitations of the different tools that may inform their decisions. This report describes five classes of models and other nonmodeling decision-support tools, then aligns those classes with the policy questions that they are best suited to address. We selectively use previous RAND Corporation work as illustrative examples, not intending to suggest that these sources represent the only or the best examples of the tools' applications. Finally, we offer nine recommendations that suggest possible ways forward to best develop and use different decision-support tools.

This research was conducted within the Forces and Resources Policy Center of the National Defense Research Institute, a federally funded research and development center sponsored by the Office of the Secretary of Defense, the Joint Staff, the Unified Combatant Commands, the Navy, the Marine Corps, the defense agencies, and the defense Intelligence Community.

For more information on the Forces and Resources Policy Center, see www.rand.org/nsrd/ndri/centers/frp or contact the director (contact information is provided on the web page).

Contents

Figure and Tables

Figure

Tables

Acknowledgments

We would like to thank the Forces and Resources Policy Center of the RAND National Defense Research Institute for allocating discretionary U.S. Department of Defense funds to support this project. We would also like to thank our colleagues Sarah Nowak of the RAND Corporation and CDR Jean-Paul Chretien of the Armed Forces Health Surveillance Center for their careful and thoughtful quality reviews of the manuscript.

Abbreviations

ANN	artificial neural network
HIV	human immunodeficiency virus
MERS	Middle East respiratory syndrome
ODE	ordinary differential equation
SARS	severe acute respiratory syndrome

Introduction

Infectious disease outbreaks present a challenge to nations and the interconnected global community. As evidenced by the outbreaks of severe acute respiratory syndrome (SARS) in 2003, influenza in 2009, Ebola and Middle East respiratory syndrome (MERS) in 2014, and the Zika virus in 2016, infectious diseases can spread rapidly within countries and across national borders. Policymakers are responsible for decisions about the nature and timing of appropriate courses of action to prevent, detect, and respond to an infectious disease outbreak. They need answers to questions about when, where, how fast, and how widely disease will spread; about the availability, effectiveness, cost, and potential unintended consequences of interventions; and about how to assess performance during and after an outbreak response. The policymakers' decisions must be made at different points as an outbreak emerges and spreads and with varying degrees of warning or available data to support those decisions. Evidence-based decision support can play a crucial role in informing these decisions, even taking critical uncertainties into account. However, modeling for decision support has not always been entirely satisfying. For example, many analyses of the 2009 influenza pandemic,[1] as well as other outbreaks,[2] have revealed failures to use models properly. Although there has been notable success and progress since then, further work is needed—for example, to clearly define expectations and to improve coordination among modelers and between modelers and clinicians, epidemiologists, and policymakers.[3]

Joseph Califano, Secretary of Health, Education, and Welfare in the late 1970s, raised a key question from the policymaker side: "How shall top lay officials, who are not [necessarily] themselves expert, deal with fundamental policy questions that are based, in part, on highly technical and complex expert knowledge?"[4] The intention of this report is to offer some guidance for how technical experts can provide appropriate information to policymakers to

---

[1] B. Y. Lee, L. A. Haidari, and M. S. Lee, "Modelling During an Emergency: The 2009 H1N1 Influenza Pandemic," *Clinical Microbiology and Infection*, Vol. 19, No. 11, November 2013.

[2] Richard E. Neustadt and Harvey V. Fineberg, *The Swine Flu Affair: Decision-Making on a Slippery Disease*, U.S. Department of Health, Education, and Welfare, 1978.

[3] Maria D. Van Kerkhove and Neil M. Ferguson, "Epidemic and Intervention Modelling—A Scientific Rationale for Policy Decisions? Lessons from the 2009 Influenza Pandemic," *Bulletin of the World Health Organization*, Vol. 90, No. 4, 2012.

[4] Joseph Califano, Jr., "Introduction," in Richard E. Neustadt and Harvey V. Fineberg, *The Swine Flu Affair: Decision-Making on a Slippery Disease*, U.S. Department of Health, Education, and Welfare, 1978.

help them make decisions on issues that are complex and that require significant technical understanding.[5]

Specifically, this report focuses on *decision support*—which encompasses the data and information that models or other approaches can provide to guide policymakers and inform decision processes[6]—related to preventing, detecting, and responding to infectious disease outbreaks. We describe models and other relevant approaches that can be used to answer policymakers' questions in this area, and we draw on previous RAND Corporation work as illustrative examples of the applications of such tools.[7] With this report, we aim to frame and respond to policymakers' information needs and consider possible ways forward to best develop and use various decision-support tools. In doing so, we hope to increase technical modelers' appreciation of the contexts in which their models are being used and broaden policymakers' understanding of the tools that may inform their real-world decisions.[8]

---

[5] These technical experts may include policymakers' own analytic staff, modelers, experts associated with the various non-modeling tools discussed later in this report, and subject-matter experts, such as infectious disease specialists.

[6] The formalization or categorization of methods for problem-solving has its own extensive literature. There are many in-depth works on this topic that are beyond the scope of this report. For an approach for policy analysis, see Eugene Bardach, *A Practical Guide for Policy Analysis: The Eightfold Path to More Effective Problem Solving*, 4th ed., Thousand Oaks, Calif.: CQ Press, 2012. For a more lighthearted and generally accessible approach to the process, see Ken Watanabe, *Problem Solving 101: A Simple Book for Smart People*, New York: Penguin Group, 2009.

[7] Our selective use of sources from RAND researchers is not intended to suggest that these reflect the only or the best examples of applications of the various models.

[8] We have aimed to provide a relevant level of technical detail, trying to avoid both too much simplification (e.g., for modelers) and too much technical complexity (e.g., for policymakers).

Decision Support: A Collaborative Endeavor

When policymakers must make decisions about uncertain events and want to understand the potential impact of different actions, they often draw on modeling and other methods created by technical experts (e.g., modelers). However, it becomes problematic when policymakers misunderstand or misuse models. This can occur because of incorrect assumptions, inappropriate use of models, failure to understand model limitations, or bad communication. Decision support involves providing evidence to guide decisions. Both policymakers (including analysts on their staff) and modelers benefit from a clear understanding of the capabilities, limitations, processes, and issues involved in constructing and using these tools. By understanding the different approaches to providing evidence-based decision support, policymakers can draw on the appropriate decision-support tools, and modelers can ensure that their contributions are useful.

Decision support should be a collaborative endeavor; ideally, modelers should understand the needs of policymakers and policymakers should understand the capabilities and limitations of modelers and their tools. In general, some models are well suited to assess the relative merits of different interventions but face challenges when required data are not available in time to make real-time decisions (rendering early data collection especially critical). In some instances, models cannot be built, modified, or run quickly enough to provide needed answers in a timely fashion. Other models are useful for rapid forecasting, even with limited data, but they cannot easily be used to compare potential interventions. Still other, nonmodeling approaches, such as exercises (i.e., gaming), can be useful for training or facilitating coordination of outbreak preparedness and response, but they are not useful for forecasting disease spread. Ideally, both policymakers and modelers should have a common understanding of this range of capabilities and limitations. In addition, modelers should have a broad understanding of the decision-support process itself and an appreciation for the information needs of policymakers and the decisions to be made.[1] This allows the modelers to suggest what aspects of the problem they are equipped to handle, when an approach or combination of approaches is warranted, and when they should defer to others.

At its best, collaboration between modelers and policymakers can make the decision-support process easier and more helpful for all parties. Policymakers provide clear statements

---

[1] While this report focuses on describing decision-support tools, there is an extensive literature on the decision-support process itself. For an excellent high-level overview for policy decisionmaking that may be useful for modelers who are unfamiliar withi it, see Bardach, 2012. For more depth about the modeling process for policy, see E. S. Quade, *Analysis for Public Decisions*, New York: Elsevier, 1975, which focuses on models that support policy. See also M. Granger Morgan and Max Henrion, *Uncertainty: A Guide to Dealing with Uncertainty in Quantitative Risk and Policy Analysis*, Cambridge, UK: Cambridge University Press, 1990, which focuses more on technical details.

of their needs and goals, as well as continuing guidance on what will or will not be useful to them. Meanwhile, technical experts tailor the tools they use and their policy advice to address these stated needs and goals, and they use a combination of methods to provide more-complete answers and ensure that the policymakers have a robust understanding of the limitations of those answers.

The Decision-Support Process

The decision-support process begins with *defining the questions* to be answered and the outcomes of interest to both the policymakers who need to answer these questions and the program managers who implement the decisions.[2] The values, criteria, and alternatives considered by policymakers must be defined at an early stage in the process (before choosing or creating the model); otherwise, as explained by Morgan and Henrion, "the result can be an analytic muddle that no amount of sophisticated analysis . . . can help to unscramble."[3] Thus, it is important to have a clear sense of the types of questions that policymakers responsible for infectious disease prevention, detection, and response might ask. Similarly, their goals must be well understood by those who will provide decision support. We categorize an illustrative list of questions in Figure 1, grouped by when they would be raised and what concerns they address.

Figure 1
Questions for Infectious Disease Policy Decisions

Disease Occurrence	How great a threat is the disease to a region, a population, or military forces?
	How likely is it that the disease will come to my country or community?
	How fast will the disease spread?
	How extensively will the disease spread?
	Across how large an area?
	When will the incidence and medical demand peak?
	How serious will an outbreak be?
	How many people will be infected? How many will die?
Planning and Preparedness	What interventions are possible?
	What is the range of intervention choices?
	What effect will interventions have?
	What are the costs and benefits or cost-effectiveness of intervention(s)?
	What intervention(s) should be undertaken?
	What effect can we realistically expect to achieve?
	How prepared are we?
	What medical capacity and capabilities are needed? How well are key (medical, emergency, public health) actors prepared and coordinated? How can we improve our preparedness? What will facilitate or impede effectiveness (e.g., timing, coverage, uptake or acceptance of intervention)?
	How cost-beneficial is preparedness?
Response	What is going on (i.e., situational awareness)?
	What medical capacities and capabilities are needed?
	How much? Where? Do we have enough of them when and where they are needed?
	How well are we doing (during a response)?
	What can we fix or do better? What can or should we replicate?
	How did we do (after a response)?
	What are the implications for next actions in this response or future responses?

---

[2] Terminology for the decision-support process differs greatly. For example, Quade (1975) calls this "initiating the analytic process" and "objectives and criteria," and Bardach (2012) calls it "defining the problem" and "select[ing] the criteria."

[3] Morgan and Henrion, 1990, p. 30.

This list will later guide the discussion between technical experts (modelers and others) and policymakers of how evidence-based decision support can be used to answer the questions.

As the figure illustrates, policymakers and program managers have a variety of concerns and many potential future decisions to make about infectious disease prevention, detection, and response. It is important to understand which types of information, coming from models or other decision-support tools, are suited for which types of decision (that is, to answer which questions).

After defining the policy questions, the next step in the decision-support process is to *select or develop models*, or *apply other techniques* appropriate to address those questions. However, doing so intelligently requires understanding existing models and techniques. The next section presents a general overview of five classes of models and nonmodeling approaches; the appendix includes more technical details on those tools.

Selecting the type of model or tool, typically the task of a technical expert (e.g., modeler or other analyst), also requires an understanding of and planning for how model outputs will be used. As Van Kerkhove and Ferguson note, "Modelling provides a means for making optimal use of the data available and for determining the type of additional information needed to address policy-relevant questions."[4] However, policymakers cannot always decipher the structure and assumptions of the models used, which was one of the problems observed during the 2009 influenza pandemic.[5] As a result, some of the most-useful models employed in outbreak preparedness and investigation are not operated in real time to inform policy decisions; instead, analysts rely heavily on classical statistical models during an outbreak to understand the data coming in.[6] This is partially a strength of statistical models,[7] but it is also a failure in preparedness to not have responsive systems in place to address these needs using other classes of models.[8]

After selecting and implementing an appropriate model or other decision-support tool, the next step in the decision-support process is to *communicate results*. Modeling is a process of approximating reality, which is limited by uncertainties about relationships and parameter values, among other factors. These limitations mean that the decisionmaking process requires both an understanding of the different types of uncertainty present in the analysis and an expertise in communicating such uncertainties.

At the most basic level, technical experts, especially modelers, distinguish between two types of uncertainty: (1) model and parameter uncertainty (that is, the model is inexact—for example, because of simplifications or uncertain assumptions) and (2) stochastic or aleatory uncertainty (that is, even if the model is correct, the predictions are inexact because of inherent randomness in the system being modeled). Modelers can describe and quantify these types of uncertainty through a variety of different means. They then must ensure that policymakers, with or without technical backgrounds, and their analytic staff understand the uncertainties

---

[4] Van Kerkhove and Ferguson, 2012, p. 307.

[5] Lee, Haidari, and Lee, 2013.

[6] Van Kerkhove and Ferguson, 2012.

[7] Martin Meltzer, "What Do Policy Makers Expect from Modelers During a Response?" presentation, CDC Grand Rounds, January 19, 2016.

[8] Modelers who provide much of the modeling resources are not necessarily available on an ad hoc basis, so preexisting networks of practitioners and responders rely only on models they can build, modify, and run themselves.

and limitations. An extensive and complex literature addresses the critical issues related to modeling uncertainties, most of which are beyond the scope of this discussion.[9]

The final step in the decision-support process is to *make and implement the decisions.*

---

[9]   For a useful introduction, see Morgan and Henrion, 1990. See also M. Brugnach, C. Pahl-Wostl, K. E. Lindenschmidt, J. A. E. B. Janssen, T. Filatova, A. Mouton, G. Holtz, P. van der Keur, and N. Gaber, *Complexity and Uncertainty: Rethinking the Modelling Activity*, Lincoln, Neb.: U.S. Environmental Protection Agency Paper 72, 2008. There is also extensive work relevant to different model types; see further discussion in the appendix.

Decision Support Using Models

As noted earlier, models constitute a large part of the decision-support tools used in public health. As Tom Frieden, director of the Centers for Disease Control and Prevention, noted, "Modeling has demonstrated its value and will continue to grow as a vital tool for public health decision making."[1] Responsible use of model-driven analysis can inform policy decisions and prevent such problems as overconfidence in models, insufficient attention to uncertainty, and lack of awareness of the difficulty of implementation.[2] Therefore, the capabilities and limitations of models and modeling should be understood before decisions are made about taking a specific approach. Investments of time and resources for building models should target their ultimate users and uses—to answer relevant questions and examine actionable alternatives. This report follows that orientation by focusing on the strengths, weaknesses, and practical applications of different types of models and, following this, discusses some of the other tools that have been, or could be, used to inform public health decisions.

While disease models can play a key role in addressing the types of policy questions discussed in Figure 1, a single model may be insufficient on its own to answer all questions, or even one question. Combinations of different model types, or of models and other decision-support tools, are often employed. Within a decision-support process, many decisions require modeling both the disease and related phenomena, such as economic impacts, or using models in conjunction with other decision-support tools, such as exercises or exploratory modeling. For example, decisions about preparedness for a disease outbreak may require a model that includes the health care system's capabilities, and decisions about the impact of a disease may require an economic model to project outcomes.

Overview of Infectious Disease Models and Their Uses

In general, a *model* is any simplified representation of reality that is intended to answer questions.[3] This general definition includes a diverse range of possible approaches. In our discussion, we limit our definition of models to techniques that represent the progression of infectious diseases in a population, all of which are mathematical or computational, although we

---

[1] Tom Frieden, "Staying Ahead of the Curve: Modeling and Public Health Decision-Making," presentation, CDC Grand Rounds, January 19, 2016.

[2] We draw from both Neustadt and Fineberg's work and more-recent work, which has developed clearer frameworks for understanding various communication failures and how they can be remedied.

[3] Quade, 1975.

note examples in which other types are used. Even within the somewhat limited scope of decision support related to infectious disease outbreaks, a wide variety of models are available. Understanding different model types and alternative approaches for supporting decisions is important for anyone intending to use them to support policy decisions and for anyone whose decisions are supported by such tools.

In this section, we establish a practical "taxonomy" of models relevant to infectious disease decisionmaking and provide a brief overview of each model type, including the basic design, typical uses, and limitations. This necessarily simplifies the complexity of how these models can overlap, be used together, or use hybrid approaches, but it should provide a reasonable initial overview.[4] In the appendix, we provide more details about how each model works, the requirements to build and use one, its limitations, and examples of how it has been used to support policy decisions related to infectious diseases.

Models of the progression of infectious diseases in a population can be divided into two broad categories. Those in the first category—*theory-based models*—use theoretical understanding of biological and social processes to represent the clinical and epidemiological course of a disease. Those in the second category—*statistical models*—bypass the details of disease and population and use statistical patterns of disease to represent disease occurrence.

Theory-Based Models

Theory-based models use scientific knowledge—drawing from biology, demography, sociology, and social psychology—to model how a disease progresses through populations and how population behavior and characteristics affect disease transmission. In general, theory-based models are more useful for examining and comparing potential interventions (some better than others within this class of model) than for forecasting future disease occurrence. This class of model requires a theoretical understanding of the pathogen, how it causes disease in a person, factors involved in disease transmission, and clinical outcomes in order to represent how much and why a disease spreads. These models therefore require an understanding of the scientific variables associated with disease spread. The detailed understanding required to build such models can be daunting, and precise models that fully represent different aspects of the disease can sometimes be impractical because scientific resources are limited. Theory-based models can be categorized into *population models*, which represent large, aggregated groups of people, and *simulation models*, which represent smaller-scale groups or individuals.

Population Models
Population models, also known as compartmental models or stock-and-flow models, divide the human population into "compartments" that represent people at different clinical stages, including susceptible (pre-infection), infected, and recovered.[5] Because population models are

---

[4] Any categorization of models is rough, but this categorization, we hope, provides a useful overview. For decision support, the model types can be, and frequently are, used in combination with one another. Given these caveats, we acknowledge that some models may not easily fit this taxonomy exactly but will share characteristics with the model types they incorporate.

[5] For a comprehensive introductory treatment, see R. M. Anderson and R. M. May, *Infectious Diseases of Humans: Dynamics and Control*, Oxford, UK: Oxford University Press, 1991.

structured around scientific parameters that capture the flow of disease transmission, they are among the best available modeling tools for understanding the dynamics of disease spread. This is partly because their requirements and complexity are less demanding than other theory-based models (described later) and partly because of their speed and flexibility. Specifically, they can be used to explore potential effects of untested interventions by comparing predicted model results assuming no intervention ("baseline") with results that introduce the intervention into model inputs. The relative simplicity in building the basic model (less-challenging data requirements, time requirements, and complexity than other types of models, including other theory-based models) and the speed and ability to represent changes in the disease dynamics make population models particularly useful for exploring different features of disease dynamics. On the other hand, the simplifications make them more limited in predictive ability, and they are less useful for modeling heterogeneous population groups. One such simplification is that simple compartmental models are deterministic and, therefore, do not capture uncertainty resulting from chance variation well. The appendix provides more technical detail and an example of using a compartmental model to understand the spread of the human immunodeficiency virus (HIV).

Simulation Models

Simulation models, broadly, represent individuals or small groups and track their status over time. Various types of simulation models use a similar modeling approach but different computational methods. They are higher fidelity than population models.[6] Simulation models expand on the scientific theory basis that underpins population models by explicitly including processes that are only approximated in an aggregated fashion in population models. The appendix discusses the advantages and disadvantages in more detail, but the obvious advantage of simulation models is their added flexibility and precision, albeit at the cost of more complexity when building the models and slower speed when running them. Simulation models can be further categorized as either *microsimulation models* or *agent-based models.*

Microsimulation models for infectious diseases represent individuals explicitly and define *events* that individuals may experience, instead of defining a fixed set of possible compartments or states the way that theory-based population models do. They use empirically derived data about disease spread to more accurately represent the spread of the disease over time, allowing a much richer, dynamic evolution of the disease spread. The empirical data can be derived from statistical regression or from other traditional statistical models using available data (discussed in the appendix).[7] For example, a microsimulation model may include a variable for how infectious an individual is, instead of assigning the individual to the category of "infectious," as in a population model, or representing the number of people with whom the person interacts, as in an agent-based model (see below). This means, for instance, that the model can represent disease progression in more detail.

Like population models, microsimulation models can be used to compare potential interventions and, to a somewhat lesser degree, to forecast disease. They are much more flexible in what they can represent than population models, because they allow understanding the

---

[6]   The methods for solving or simulating the different types of models can differ greatly, but for our purposes, these are computational, rather than modeling, concerns.

[7]   For a more complete overview of these models in a variety of areas, see Cathal O'Donoghue, *Handbook of Microsimulation Modelling*, Bingley, UK: Emerald Group Publishing, 2014.

interactions between different parts of an entire policy system, including representation of medical care, economic factors, and many other important features for modeling complex interventions. Compared with population models, however, microsimulation models require more data, a more complex build process, and much more computational work to represent individuals (versus the aggregated populations represented in a population model).

Agent-based simulation models describe the interplay between the *behavior* of individuals and their disease status. While these models can be considered a special case of microsimulation models, the typical focus of agent-based models is different. They represent the behaviors of individuals instead of the types of empirically derived data about disease spread used by microsimulation models. Individuals' behaviors follow predefined rules and can be functions of different endogenous factors. For example, people may avoid other people when they observe that many others are infected, thereby slowing the spread of disease. Similarly, they may be more likely to seek immunization if they see that many people were infected the previous year. This can create "emergent" properties of a system—those that are caused by the combination of factors—and these properties evolve over time. This model capability allows complex disease dynamics to result from relatively simple rules. Representations of the individual behaviors can lead to complex patterns of disease spread, and the resulting simulation can be analyzed to track the spread of disease that emerged from individuals' actions. As a consequence, agent-based models can show when rapidly spreading diseases may be less damaging as a result of behavioral changes that blunt the spread of the disease, or they may exhibit complex multiyear patterns.

Agent-based models are well suited for comparing potential interventions and not as well suited for forecasting disease. They require several different types of inputs, more than most other model types, and are generally more expensive to create and evaluate than population models. Data requirements are moderate—more than for population models and different from the empirical data for microsimulation models. Similar to that of microsimulation models, the complexity of agent-based models is also greater and the computational speed slower than other model types described in this report.

One example of a policy application of an agent-based model comes from a project led by one of the authors of this report (Raffaele Vardavas) on behaviors toward vaccination based on different features of individuals' social networks.[8] Such modeling can be used to determine the extent to which interventions to influence social networks would be successful. The appendix provides further details of this illustrative example.

Statistical Models

Statistical models are distinct from theory-based models. They use mathematical relationships to directly represent quantities of interest, such as using past observed data to forecast future events (e.g., disease occurrence), relying on mining large amounts of data. Statistical models can reproduce dynamic real-world relationships by learning trends from empirical data and encoding these dependencies in a mathematical model, without directly representing the causal

---

[8]   Raffaele Vardavas and Christopher Steven Marcum, "Modeling Influenza Vaccination Behaviour Via Inductive Reasoning Games," in Piero Manfredi and Alberto D'Onofrio, eds., *Modeling the Interplay Between Human Behavior and the Spread of Infectious Disease*, New York: Springer-Verlag, 2013.

scientific factors involved. These models use real-world outputs (i.e., the empirical "training" set of observed data), which allows the statistical method to learn the correct behavior (e.g., disease occurrence).

In other words, the learned mathematical relationships, or patterns, are applied to forecast future disease occurrence. Statistical models of all types enable this data-driven, inductive prediction under limited model assumptions, despite potentially limited exact theoretical understanding. Because of this, practically all inductive learning includes an implicit assumption of *invariance*,[9] meaning that the past relationship between the disease characteristics and occurrence continues to tell us about future outbreaks.

We can take this further in the context of decision support for infectious diseases: Models should be used to inform decisions only in scenarios similar to the model-building context; to the extent that the scenarios differ, any inference is less valid. For example, if a model uses data from only one location, or with just one type or intensity of intervention, using that model to make similar forecasts in a different location or under different conditions can be problematic.

Statistical models are especially susceptible to this kind of problem because of their strong dependence on the training data. Additionally, there is a related, but more fundamental, problem specific to typical applications of statistical models. This is the problem of inferring causality: How can we infer a causal link between two events using statistical methods? This is often critical for questions of the relative merits of interventions (alternative pasts or future forecasts). Statistical models used for forecasting (as opposed to econometric statistical models used for evaluation) have difficulty answering such questions.

We describe two general classes of statistical models for representing diseases—traditional *regression-based models* and modern *machine-learning models*.[10]

Regression-Based Models

Regression-based models have been extensively used to model the real-world course of a disease and consequently can be considered a standard or traditional statistical approach.[11] Through mathematical formulas, they seek to produce an optimal fit between input variables (potential predictors) and observed outcomes. These learned mathematical relationships do not require specification or even understanding of underlying theory—for example, related to spread of a specific pathogen. They are predictive models, not explanations, so they may not tell the full story. Regression-based statistical methods can also be limited in the range of relationships that they can model. Despite these limitations, such models are very computationally fast and can be used to model and understand relationships in the data even when the scientific relationships are unknown. This means that quick and roughly accurate models can predict outcomes even when the causes are not known or understood, or when the scientific theory is insufficient for representation.

---

[9]   Leslie Valiant uses this term, which, in our setting, says that the dynamics and natural processes behind the disease being modeled do not change radically. See Leslie Valiant, *Probably Approximately Correct: Nature's Algorithms for Learning and Prospering in a Complex World*, New York: Basic Books, 2013.

[10]   These categories overlap in many ways, and our explanation is again simplified. Many machine-learning models are generalizations (or simplifications) of traditional models, and so-called traditional regression-based models are able to represent much more-complex dynamics than our simple examples might otherwise suggest.

[11]   We exclude from our discussion the use of these models to evaluate randomized clinical trials or trial interventions, because those are applications for evaluating an intervention, not modeling the disease progression.

Machine-Learning Models

Machine-learning models examine observed input and output data and use algorithmic statistical techniques to mathematically "learn" structured relationships in the data. Modern machine-learning methods extend the expressive range of traditional regression-based statistical methods to include more-complex dynamics. They can use more-complex adaptive mathematical structures to represent arbitrarily complex relationships. Therefore, their usefulness for disease-forecasting decision support is unsurprising. The basic approach is to use sample input-output data to learn complex relationships with the goal, in this context, of predicting future outputs. The complexity of the relationships makes machine-learning models more predictive but often harder to understand or explain. There are several powerful, versatile, and robust families of such models (see the appendix for more details).

Comparison of Theory-Based and Statistical Models and Their Limitations

As a very high-level overview, we now compare the key characteristics of the three theory-based and two statistical models described earlier (see Table 1). The relative utility and degree of challenge for each model type are based on the judgment of the project team and are intended as a rough overview, not an authoritative conclusion. The information in the table represents the typical case of each model, and specific techniques that exist for building different models can change the relative rankings. As discussed, theory-based models differ from statistical models in the way that they handle causality: Theory-based models incorporate causality; statistical models do not.[12] Model specifications in theory-based models encode

Table 1
Comparison of Model Characteristics

Characteristic	Theory-Based Models			Statistical Models	
	Population	Micro-simulation	Agent-Based Simulation	Regression Based	Machine Learning
Use[a]					
Forecast disease	light	light	dark	light	dark
Compare interventions	dark	medium	medium	dark	dark
Challenge[b]					
Data	dark	dark	light	light	dark
Theory	light	light	dark	medium	dark
Time (speed)	dark	dark	dark	dark	light
Modeling complexity	light	light	dark	light	light
Communicating results	dark	light	light	medium	dark

[a] Scale for use: Dark green = very useful; light green = useful; yellow = somewhat or possibly useful; red = typically difficult or not used.

[b] Scale for challenge: Dark green = least challenging; light green = only somewhat challenging; yellow = challenging; red = very challenging or most challenging.

---

[12] D. Lewis, "Causation," *Journal of Philosophy*, 1973, pp. 556–567; J. Pearl, *Causality*, Cambridge, UK: Cambridge University Press, 2009.

causal chains, or "laws," for the modeled phenomena. In population models, for example, the coupled differential equations will specify scientific facts, such as "an increase in the susceptible population causes an increase in the rate of infection." We can infer an explanation of observed model outputs directly from the model. Such explicit specifications (when they are accurate) allow for robust simulations of what-if or counterfactual scenarios. Such scenarios are important for informing decisions about interventions. Statistical models do not naturally distinguish events in a causal chain from events that are merely correlated. Inferring causality using statistical models on observational data requires extra machinery.[13] James Heckman,[14] Judea Pearl,[15] Paul Rosenbaum and Donald Rubin,[16] and others[17] have proposed alternative approaches to causal inference.[18]

As a general overview, theory-based models work well when there is sufficient understanding and information on the disease dynamics to inform models meaningfully and when the models are sufficiently validated in the context considered. They are generally excellent for comparing potential interventions but less useful than statistical models for forecasting disease incidence. Of the five models compared in Table 1, population models are a good approach when used for qualitative understanding of disease outbreaks. In particular, they are excellent for comparing potential future interventions. They also run rapidly, they require less input data, and their results and structure are most easily communicated. However, they are harder to tune to match observed outbreak statistics or to give accurate quantitative predictions. On the other hand, regression-based and machine-learning statistical models can match observed statistics and give more-accurate forecasts without a full accounting of underlying disease characteristics. But they require considerably more data to be useful.

In general, theory-based models provide a good qualitative description of how outbreaks evolve and what their dependencies are. Because of the models' generality and requirements for underlying scientific information, they can be difficult to train for accurate prediction with limited resources. Within the larger group of theory-based models, population models provide a moderately useful level of understanding across a range of problems both rapidly and at low cost, but their limited accuracy can prove problematic. Simulation models can

[13] Identifying causal links between events involves building support for two related hypotheses: the conditional hypothesis ("If A, then B"—for example, if the susceptible population increases, then the rate of infection increases) and its counterfactual hypothesis ("If not A, then not B"). Observational data (i.e., data recording observed events) can provide statistical support only for the conditional hypothesis because it accounts only for observed scenarios, not unobserved scenarios. This amounts to establishing correlations between observed events. Accordingly, observational data cannot directly support counterfactual hypotheses because counterfactuals involve unobserved, nonoccurring events. The goal of experimental design for randomized controlled trials is to observe data that address counterfactuals. Such trials remain the gold-standard approach to causal inference in spite of more-recent advances in quasi-experimental methods.

[14] J. J. Heckman, "Econometric Causality," *International Statistical Review*, Vol. 76, No. 1, 2008.

[15] Pearl, 2009.

[16] P. R. Rosenbaum and D. B. Rubin, "The Central Role of the Propensity Score in Observational Studies for Causal Effects," *Biometrika*, Vol. 70, No. 1, 1983.

[17] L. Bottou, J. Peters, J. Quinonero-Candela, D. X. Charles, D. M. Chickering, E. Portugaly, D. Ray, P. Simard, and E. Snelson, "Counterfactual Reasoning and Learning Systems: The Example of Computational Advertising," *Journal of Machine Learning Research*, Vol. 14, No. 1, 2013.

[18] The methods of Heckman and Pearl are based on structural-equation models, directed acyclic graphs, and special operators for evaluating interventions. Rubin avoids structural-equation models and simulates randomized controlled trial conditions using a family of matching methods. This allows for some limited evaluation of treatments or interventions.

address some of these limitations (especially for understanding interventions) but not all of them, and the models are slower and more expensive to construct. They are constrained by uncertainty, limited knowledge, and the difficulty of fully modeling the scientific processes involved in disease spread.

Statistical models can make useful inferences and predictions when we do not have a full understanding of the constituent processes of the phenomenon of interest—that is, data inputs do not require an understanding of underlying causal mechanisms. But the use of statistical models comes at a price: higher sensitivity to violations of the invariance assumption (which is especially true when modeling complex phenomena with limited data[19]) and difficulty representing the underlying causal relationships. Within the larger group of statistical models, traditional regression-based statistical models are easy to explain and are used widely to understand relationships between individual variables for use in other models, as well as for rapid and simple approximations of trends or relationships. In contrast to this, machine-learning models are most useful for forecasting disease (i.e., prediction before an event), although, as noted earlier, data limitations are critical. Such models are also typically harder to explain to a nontechnical audience because the basis is mathematical instead of related to the origins of the disease. Both machine-learning and regression-based statistical models depend heavily on sufficient data; thus, additional data sources can make them more powerful, more useful, and applicable to more purposes. They can be used for intervention analysis, although only in cases in which there is a history of the intervention with sufficient high-quality data, which is rare given the significant data requirements.

Nonmodeling approaches (which are described in the next section) are typically more valuable for real-time decisions. They are informed by modeling, which is best undertaken before real-time decisions are needed. Most modeling processes outside of government agencies are not well adapted to this type of work.

Theory-based models and statistical models are sensitive to different types of errors. Consider, for example, factors commonly responsible for inaccuracies in models of real-world phenomena, such as disease incidence. One factor is the use of incorrect or improper parameters in models. The use of more data, more support from experts, and robust optimization methods can address this factor. Another factor is the use of models constructed under an imperfect or incomplete understanding of the causative links producing the phenomenon. This problem, termed *model inadequacy*, encapsulates the idea that most models are just approximations of real dynamics, and some approximations are too rough to capture dynamics of interest.

Alleviating the model-inadequacy problem requires refining the model with further scientific understanding or using more-expressive mathematics, but this solution assumes that there is relevant information or data to augment the model. Thus, model inadequacy can be tough to address in theory-based models. However, statistical models provide an alternative approach to addressing the problem. The supervision or machine-learning approach makes the effects of model inadequacy directly diagnosable in training. The downside is that final-tuned, data-driven statistical models are often difficult to interpret.

---

[19] Complex phenomena demand complex models, and complex models have large numbers of parameters. The number of data samples needed to tune the model increases exponentially with the number of model parameters (the curse of dimensionality; see the appendix).

Nonmodeling Decision-Support Approaches

While models are useful heuristic tools to support policy decisionmaking, they are not the only tools in a policymaker's repertoire. Outside of mathematical modeling, there are many useful tools for structuring decisions, simulating decision contexts, and synthesizing expertise for systematic application. These require different types of technical experts (that is, typically not modelers) but are also complex and need to be used carefully. Below, we outline several approaches or tools that are relevant to policy questions about infectious disease prevention, detection, and response.

One of the most widely used nonmodeling approaches is *public health surveillance*, which refers to "the ongoing systematic collection, analysis, interpretation, and dissemination of data regarding a health-related event for use in public health action to reduce morbidity and mortality and to improve health,"[1] or, more simply, "systematic information for public health action."[2] Public health surveillance monitors disease occurrence to discern trends and detect anomalies, including rare events. The design and capabilities of a particular surveillance system determine the likelihood and timeliness of detecting a disease and the accuracy of reporting.[3]

The *expert elicitation* approach uses subject-matter expertise drawn from those with a deep understanding of the problem at hand, either to directly advise on a decision or to estimate particular inputs or values that can be used in decisionmaking. As a method, expert elicitation is capable of avoiding many pitfalls that can accompany the use of complex models, but it has significant drawbacks as well.

As explained by Sutherland and Burgman, "The accuracy and reliability of expert opinions [are] compromised by a long list of cognitive frailties," but structured methods "alleviate the effects" of biases and the overreliance on individual opinion.[4] For example, the Delphi method developed at the RAND Corporation in the 1950s,[5] RAND's online ExpertLens™ tool,[6] and similar methods are capable of systematically capturing and then utilizing expert

---

[1]   S. Thacker, "Historical Development," in Steven Teutsch and R. Elliot Churchill, eds., *Principles and Practice of Public Health Surveillance*, New York: Oxford University Press, 2000.

[2]   Melinda Moore, Edward Chan, Nicole Lurie, Agnes Gereben Schaefer, Danielle M. Varda, and John A. Zambrano, "Strategies to Improve Global Influenza Surveillance: A Decision Tool for Policymakers," *BMC Public Health*, Vol. 8, 2008.

[3]   Moore et al., 2008.

[4]   William J. Sutherland and Mark Burgman, "Policy Advice: Use Experts Wisely," *Nature*, Vol. 536, October 2015, p. 317.

[5]   RAND Corporation, "Delphi Method," web page, undated c.

[6]   S. Dalal, D. Khodyakov, R. Srinivasan, S. Straus, and J. Adams, "ExpertLens: A System for Eliciting Opinions from a Large Pool of Non-Collocated Experts with Diverse Knowledge," *Technological Forecasting and Social Change*, Vol. 78, No. 8, 2011.

opinion while mitigating these biases. One example of the modified Delphi process in use for public health emergencies was conducted in the mid-2000s. It was motivated by recent experiences with the West Nile virus and SARS, as well as preparedness for potential outbreaks, such as monkeypox. This expert elicitation focused on evaluating a logic model and associated questionnaire designed to measure preparedness for infectious disease outbreak decisionmaking in the United States. The resulting 37-item performance measurement tool was found to reliably measure public health functional capabilities in a tabletop exercise setting, with preliminary evidence of a factor structure consistent with the original conceptualization and of criterion-related validity.[7]

Exercises, also sometimes called *games* or *gaming*, include a variety of activities, such as tabletop exercises and drills. These activities place a range of key actors, from high-level policymakers to implementers on the ground, in a simulated emergency. The intensity of the exercise depends largely on the stage of planning (early or advanced) and the resources available.

Exercises can range from a meeting with structured content intended to elicit thoughtful decisionmaking through conversation (that is, a tabletop exercise) to small-scale operational drills or full-scale simulated emergencies in which actors play victims and actual responders must participate. Exercises are carefully designed to target particular decisionmaking audiences (e.g., policymakers, hospital administrators, emergency room physicians, military logistics officers) and can be used to inform their planning and future decisions, elicit concerns in a structured way, find and explore alternatives that were not previously considered, or train participants. For example, between 2004 and 2007, RAND researchers designed and conducted several tabletop exercises on pandemic influenza preparedness, both in the United States and internationally. At the state level, this was done to understand capabilities and to train participants,[8] while, at the international level, similar exercises were used to practice, test, and evaluate responses to facilitate further planning.[9]

Policy analysis frameworks use best practices from policy analysis research to describe and consider the merits of a potential decision or intervention or to compare alternative interventions. These frameworks can help policymakers to answer some of the key questions that arise quickly with an infectious disease threat and to assess the value of new or existing interventions for current and future threats. They can answer a question about a particular intervention (is it a good idea?) or help policymakers compare alternative interventions (which one is best?). An example of such a policy analysis framework was developed by RAND researchers as a proof of concept in response to the 2014 Ebola outbreak.[10] This tool is flexible enough to allow evaluation of a single intervention, a few interventions with the same aim, or an entire landscape of

---

[7]   E. Savoia, M. A. Testa, P. D. Biddinger, R. O. Cadigan, H. Koh, P. Campbell, and M. A. Stoto, "Assessing Public Health Capabilities During Emergency Preparedness Tabletop Exercises: Reliability and Validity of a Measurement Tool," *Public Health Reports*, Vol. 124, No. 1, 2009.

[8]   D. J. Dausey, J. W. Buehler, and N. Lurie, "Designing and Conducting Tabletop Exercises to Assess Public Health Preparedness for Manmade and Naturally Occurring Biological Threats," *BMC Public Health*, Vol. 7, 2007; and N. Lurie, D. J. Dausey, T. Knighton, M. Moore, S. Zakowsky, and L. Deyton, "Community Planning for Pandemic Influenza: Lessons from the VA Health Care System," *Disaster Medicine and Public Health Preparedness*, Vol. 2, No. 4, 2008.

[9]   D. J. Dausey and M. Moore, "Using Exercises to Improve Public Health Preparedness in Asia, the Middle East and Africa," *BMC Research Notes*, Vol. 7, 2014.

[10] Margaret Chamberlin, Shira Efron, and Melinda Moore, *A Simple Approach to Assessing Potential Health Emergency Interventions: A Proof of Concept and Illustrative Application to the 2014–2015 Ebola Crisis*, Santa Monica, Calif.: RAND Corporation, PE-148-RC, 2015.

interventions. It defines several criteria against which an intervention can be judged—efficacy, ease of intervention, cost, risk of unintended consequences, political viability, social and cultural viability, equity, and time frame. It then provides a real-world example of a single intervention and an example that compares two alternative interventions, populating data for each criterion based on information from published papers and the media and color-coding each criterion using a simple green/yellow/red scheme. This illustrative example provides a simple, practical, proof-of-concept policy analysis tool that aims to fill potential gaps in a policymaker's ability to systematically assess potential interventions.

Even though the illustrative application of that tool is focused on Ebola, the conceptual approach and types of decisions are not unique to this public health emergency. Tools of this kind could be particularly useful in planning for or making decisions during response to different kinds of disasters, including both naturally occurring and manmade. In addition, the tools are flexible enough to accommodate policymakers' time constraints: They could be successfully used to help policymakers decide quickly when needed on the basis of a swift review of published reports and consultation with just a few people in a short meeting, or they could be applied more rigorously using in-depth data collection, consultation, and analysis, if time permits.

An *intra-action report* is a practical, visually simple proof-of-concept tool also developed by RAND researchers in response to the 2014 Ebola outbreaks.[11] It provides a systematic way to capture and communicate progress over the course of an emergency response to inform modifications to that response. This is a useful addition to traditional after-action reports, which are assessments of disaster response and recovery that are typically undertaken after an event to inform response to a future emergency. After-action reports typically focus on cataloging failures, often not capturing things that were done well. The intra-action report makes a clear distinction between simply documenting a success or failure and capturing a "lesson learned"—which is when action is taken to mitigate a problem or replicate a success. This tool provides a framework to track, synthesize, evaluate, and communicate the problems and successes identified or the lessons that are being learned during an ongoing public health emergency or disaster response and recovery effort and apply them during that same effort. An intra-action report also enables the capture of actions to successfully address initial negative experiences (i.e., lessons learned from initially doing things wrong) and replication of initial successes (i.e., lessons learned from doing things right—the "positive deviance" approach[12]).

---

[11] Margaret Chamberlin, Adeyemi Okunogbe, Melinda Moore, and Mahshid Abir, *Intra-Action Report—A Dynamic Tool for Emergency Managers and Policymakers: A Proof of Concept and Illustrative Application to the 2014–2015 Ebola Crisis*, Santa Monica, Calif.: RAND Corporation, PE-147-RC, 2015.

[12] Richard T. Pascale, Jerry Sternin, and Monique Sternin, *The Power of Positive Deviance: How Unlikely Innovators Solve the World's Toughest Problems*, Boston: Harvard Business Press, 2010.

Alignment Between Policy Questions and Decision-Support Approaches

Now that we have described several modeling and nonmodeling approaches, we can begin to align these with the expected policy questions (see Figure 1) that the approaches can support. Table 2 summarizes this alignment. Similar to in Table 1, the ratings here are based on our judgment and are intended as a rough overview, not an authoritative conclusion. In this case, they also assume sufficient theoretical understanding to create population models, as well as sufficient data to train the statistical models.

In general, the theory-based models excel at addressing questions about understanding the phenomenon being modeled (e.g., What interventions are possible? How great is the threat?). The statistical models excel at mechanistic quantitative predictions concerning the phenomenon of interest (e.g., How fast will it spread? How many resources do we need to respond?). Both modeling approaches benefit greatly from data obtained from public health surveillance systems. Such data sources help anchor the models to the realities on the ground.

In addition to their standard functions, policy analysis frameworks and intra-action reports can contribute to modeling—for example, to help shape the basic questions that the modeler should address. Using such tools to support modeling ensures that the results of modeling efforts can lead more naturally to actionable recommendations. The theory-based models can be useful for analyzing potential interventions in this context. As described in the appendix, developing hybrids of theory-based and statistical methods can improve the estimation accuracy of hypothetical intervention effects (e.g., to give high-accuracy estimates of the relative effects of interventions A and B).

Table 2
Applicability of Models and Nonmodeling Approaches to Policy Questions

Questions for Infectious Disease Policy Decisions	Theory-Based Models			Statistical Models		Nonmodeling Approaches
	Population	Micro-simulation	Agent-Based Simulation	Regression Based	Machine Learning	
Disease Occurrence						
How great a threat is the disease to a region, a population, or military forces?						Public health surveillance, expert elicitation (combined with modeling)
How fast will the disease spread?	*					Public health surveillance
How extensively will the disease spread?						Public health surveillance
When will the incidence and medical demand peak?						Expert elicitation (e.g., for model assumptions)
How serious will an outbreak be?						Expert elicitation
Planning and Preparedness						
What interventions are possible?						Expert elicitation
What effect will interventions have?						Review of published literature
What are the costs and benefits or cost-effectiveness of intervention(s)?						Review of published literature
What interventions should be undertaken?						Economic analyses (cost-benefit, cost-effectiveness)
How prepared are we?						Best-practice documentation; policy guidelines
How cost-beneficial is preparedness?						Gaming, tabletop or full-scale exercises, drills
Response						
What is going on (i.e., situational awareness)?						Cost-benefit analysis
What medical capacities and capabilities are needed?						Public health surveillance (if suited for real-time data)
How well are we doing (during a response)?				**		Local, state, and federal emergency operations centers, including Internet-based resource-monitoring tools
How well did we do (after a response)?				***		Intra-action report
						After-action report

a Scale for applicability: Dark green = very applicable; light green = applicable; yellow = somewhat or possibly applicable; white = not typically applicable.

* Population modeling could be used to estimate R_0, for example. See the appendix for more information on R_0.

** Traditional regression-based modeling could be used for understanding real-time data, for example.

*** Traditional regression-based modeling could be used for econometric evaluation, for example.

Recommendations and Discussion

To enable better use of decision-support tools to inform infectious disease prevention, detection, and response, we offer several recommendations. These recommendations derive from the literature that we reviewed and our assessment of the characteristics and applicability of different types of models and nonmodeling approaches. They also reflect earlier discussion of the decision-support process, the ideally collaborative nature of decision support (i.e., collaboration between policymakers and modelers or other technical experts), and the roles of each side. We organize the first eight recommendations based on whose action is required and then conclude with a ninth recommendation about the modeling ecosystem as a whole. The discussion that follows elaborates on each of these recommendations.

For *modelers and policymakers*, collectively:

1. Establish partnerships and communications with one another before an emergency arises.
2. Coordinate at multiple levels, including with program managers responsible for implementing actions and interventions.
3. Ensure the timely availability of data needed for modeling.

For *policymakers* (including their analytic staff):

4. Clarify priority policy questions and ensure that modelers understand them.
5. Use relevant nonmodeling approaches to provide decision support.

For *modelers or policymakers' analytic staff*:

6. Use the most-appropriate models and other approaches to meet specific real-world needs.
7. Set appropriate expectations for the use of models.

For *modelers*:

8. Improve models and modeling approaches.

For organizations involved in policymaking and technical support, including modeling:

9. Improve the modeling ecosystem.

Establish partnerships and communications before an emergency arises. We have emphasized that the decision-support process is a collaborative endeavor between policymakers (who might not be technical experts) and the technical experts on whom they rely for decision support. In the interest of time and efficiency, it is important to have partnerships and bureaucratic structures in place before an emergency arises. This can "be facilitated by the establishment of national and international modelling networks such as those that were created in 2009."[1] These networks can then be used for planning beforehand and then for sharing data and collaborating on responses as planned once an emergency arises. Modelers and other technical experts are found in academic institutions, other research institutions, government, and the private sector. Those who will be most useful for supporting public-sector policy decisions understand the policy context and needs of policymakers and can select the most-appropriate model(s) to use for a specific question or set of questions.

As the influenza pandemic emerged in the spring of 2009, officials within federal government and state and local health departments used a variety of data types to inform key decisions. In the aftermath of the pandemic, a symposium was held to evaluate how effectively the data derived from a variety of biosurveillance sources were used for analysis and decisionmaking during the event.[2] Although generally successful, data-sharing was ad hoc and not utilized fully; as a result, symposium participants identified improved partnerships as a key need going forward. To address this need, health agencies that have used particular modeling institutions opportunistically in the past can leverage them in the future, both for sharing data and for coordinating action. In addition, agencies that have not previously leveraged modeling or decision support can work to set up partnerships in advance of an emergency to better facilitate their use in the future.

We echo several past authors in emphasizing the critical importance of honest reciprocal communication between policymakers and all relevant technical experts, such as modelers, data managers, evaluation experts, program managers, and infectious disease specialists. Not only must expectations from and limitations of modeling be clear, but data should be shared collaboratively once available, and the preliminary modeling results should be made available rapidly to policymakers and others who may need to act on results, such as program managers in the field.[3] If the uses for models can be specified or planned in advance, this can provide an excellent opportunity for trial runs and ensuring that communications are clear.

Coordinate at multiple levels. Any coordination between technical experts and policymakers should involve all relevant parties. Depending on the context, this may include U.S. and foreign government policymakers (local, state, or national), technical experts in research institutions (academic and other), nongovernmental organizations, the private sector, and any others with a stake in the issues and decisions to be addressed. Even though international cooperation is not new, there are significant disparities across countries and at various levels of government in the ability to provide data, perform modeling, and use the results for plan-

---

[1] Van Kerkhove and Ferguson, 2012, p. 308.

[2] M. Lipsitch, L. Finelli, R. T. Heffernan, G. M. Leung, and S. C. Redd, "Improving the Evidence Base for Decision Making During a Pandemic: The Example of 2009 Influenza A/H1N1," *Biosecurity and Bioterrorism: Biodefense Strategy, Practice, and Science*, Vol. 9, No. 2, June 2011.

[3] At times, there is a conflict between policymakers, who expect data (or modeling outputs) to be quickly available, and modelers or other technical experts, who may be reluctant to share preliminary results that they consider to be insufficiently validated.

ning. Communication of limitations and uncertainties is even more difficult across these diverse groups.

Ensure the timely availability of data needed for modeling. Modeling always benefits from having more data available when needed, especially when the data are high quality and well documented.[4] It is important to keep in mind that "modeling cannot substitute for data,"[5] and modelers typically are not the source or owners of data used in their models. All model types rely on some amount of data, and, to the extent that the model can be built knowing what data will be available during a crisis, the model itself can be better adapted for the needs of policymakers. This requires both planning before a crisis and collaboration during it.

During the 2009 influenza pandemic, appropriate real-time data were not always readily available despite being collected, and, thus, modeling results were not generated, shared, or disseminated in a timely manner.[6] This report has described that various kinds of models differ in the amount and types of data needed. If modeling infrastructure is established in advance of an emergency (for example, if influenza modeling efforts are planned in advance to incorporate anticipated biosurveillance and other data), modelers and policymakers can determine the nature and timeliness of data that are needed for the most-appropriate model types. Policymakers can then ensure that these data will be available for modeling before and as an emergency unfolds, allowing the modelers to generate, validate, and share results more quickly.

Clarify the priority policy questions and ensure that modelers understand them. As part of the partnerships with modelers, it is helpful for policymakers to articulate the questions for which they seek answers from models. Despite the fact that we are unsure exactly what will occur, and when, there are a variety of questions that policymakers know will need to be answered. It is equally important for modelers to have a clear understanding of the real-world decisions that the policymakers they are supporting will need to make. This enables selection of the right technical experts to provide the most-appropriate decision support. In some instances, models can be built before an outbreak emerges, or the needs can at least be discussed. Ideally, to ensure that communication is clear, the dialogue between modelers and policymakers should start well in advance of the decision process, not after an infectious threat emerges.

Use relevant nonmodeling approaches to provide decision support. In this report, we have described several useful nonmodeling approaches and tools that can help answer policy questions. Different approaches are pertinent to different sets of questions. For example, public health surveillance and expert elicitation can help us understand disease occurrence. A literature review and exercises or gaming can contribute to preparedness and system improvements. Public health surveillance, intra-action reports, and after-action reports can help address policy questions during and following an outbreak response.

Use the most-appropriate models and other approaches to meet specific real-world needs. As we have emphasized throughout this report, each type of tool has its own unique profile of capabilities, limitations, and applicability. In addition, combining models together or with nonmodeling approaches is often appropriate. Our policy-oriented approach has aimed to place modeling within the context of real-world questions that need real-world decisions and actions.

---

[4] Lauren Ancel Meyers, "Modeling to Support Outbreak Preparedness, Surveillance and Response," presentation, CDC Grand Rounds, January 19, 2016.

[5] Van Kerkhove and Ferguson, 2012.

[6] Lee, Haidari, and Lee, 2013.

Modelers deploy their technical expertise to produce models that can usefully inform such decisions. A critical part of that process is to select the most appropriate model, nonmodeling approach, or combination for the policy questions at hand.

Set appropriate expectations for the use of models. It is important for modelers to understand and clearly communicate the capabilities and limitations of each type of model and the policy questions that models can help address.[7] We have emphasized the importance of understanding these details, and we describe several examples in this report (see the appendix for more detail). This level of understanding of different models and tools and their alignment with policy questions can help policymakers and technical experts select the most-appropriate tools to inform critical policy decisions in the future. This, in turn, can help set realistic expectations and avoid the problem of unmet expectations during the "fog of war."

Improve models and modeling approaches. Models and modeling approaches can be improved to some degree to address policy questions. Once a decision-support system is defined, it is possible to understand what key limitations exist and the relative importance of alleviating them. Some model improvement can be achieved through investment in computational tools and better understanding and representation of diseases; this is important ongoing work, although it falls well outside the immediate scope of most policymakers' work. And because much work has already been done, attempts to improve models are subject to diminishing returns. In contrast, improving processes and protocols for using such models and providing improved support are often within the control of policymakers—and provide larger potential gains.

Improve the modeling ecosystem. Disparate systems used to inform policymakers often end up providing confusion instead of insight, and complex models are ignored if the implications are unclear. To ameliorate this problem, specific groups or centers can be formed to interpret information and model results from various teams and data sources and to ensure that these teams all have access to the appropriate data. With sufficient funding and a change to how modeling is used in government, this could resemble the way in which the National Oceanic and Atmospheric Administration runs the National Hurricane Center, which provides resources for predictions of hurricanes and functions as the key provider of hurricane data. While ambitious, a properly considered and well-implemented center for infectious diseases and modeling would be able to coordinate and implement many of the steps recommended here. Additionally, this type of infrastructure could coordinate to ensure that policymakers understand the implications of the complex models being used and that modelers understand the ways in which the models will support policymaker decisions.

---

[7] It is especially important to communicate limitations because those limitations may have policy solutions. Significant future uncertainty is a modeling drawback, but it also strengthens the argument for more-robust, rather than more-prescriptive, approaches in policymaking.

Conclusions

This report seeks to fill a gap in the decision-support literature by raising and answering key questions to inform the best use of models and nonmodeling approaches to answer relevant real-world policy questions. With the continuing threats posed by infectious diseases worldwide, it is important to understand the models and other approaches available to inform decisions about prevention, detection, and response and to use these decision-support tools appropriately. As suggested by others,[1] there is a need for further coordination with data, models, and planning. We can and should learn from recent experiences, such as the 2009 influenza pandemic, Ebola and MERS in 2014, and the Zika virus in 2016. Review of such past experiences and the recommendations described in this report suggest opportunities to make optimal use of these tools.

---

[1] See, for example, Van Kerkhove and Ferguson, 2012.

An In-Depth Look at Theory-Based and Statistical Models

Theory-Based Models

As described in the report, *theory-based models*,[1] also known as mathematical models,[2] use scientific knowledge to model how a disease progresses through populations and how population behavior and characteristics affect disease transmission. This class of model requires a theoretical understanding of the pathogen, how it causes disease in a person, factors involved in disease transmission, and clinical outcomes in order to represent how and why a disease spreads. These models therefore require an understanding of the scientific variables associated with disease spread. Theory-based models can be categorized into *population models*, which represent large, aggregated groups of people, and *simulation models*, which represent smaller-scale groups or individuals.

Population Models

Population models, also known as compartmental models or stock and flow models, divide the human population into "compartments" that represent different clinical stages, including pre-infection, infection, and recovery.[3] As a simple example, "SIR" models represent everyone as initially belonging to one of three populations: susceptible, infected, or recovered (the last of these is considered immune from further infection). The subpopulations, or compartments, represent the flow of disease transmission through a population. The aggregated totals in each subpopulation evolve over time as determined by the model, so that susceptible people exposed to the disease get infected, then recover. More-complex variants of this type of model can have compartments representing more-detailed breakdowns of the disease progression as well, such as "SEIR" models (which include a stage of exposure before infection); can represent multiple strains of a disease agent in a population (e.g., multiple strains of the influenza virus or dengue

---

[1] The *theory* in *theory-based models* refers to the underlying theory of the scientific disciplines relevant to the phenomena being modeled—in this instance, infectious diseases. It does not refer to the theory underlying the formulation, training, and representative power of the models. All models have significant theory behind their structure. The theories underlying theory-based infectious disease models are primarily causal and structural laws established in the natural sciences. In contrast, the theory behind statistical models is concerned primarily with the representation and algorithmic learning of observed natural behavior. Similarly, operational exercises and other nonmodel-based approaches described in this report also have significant theory behind them.

[2] M. Choisy, P. Sitboulang, M. Vongpanhya, C. Saiyavong, B. Khamphaphongphanh, B. Phommasack, F. Quet, Y. Buisson, J.-D. Zucker, and W. van Pahuis, "Rescuing Public Health," in Serge Morand, Jean-Pierre Dujardin, Régine Lefait-Robin, and Chamnarn Apiwathnasorn, eds., *Socio-Ecological Dimensions of Infectious Diseases in Southeast Asia*, Springer-Verlag, 2015.

[3] For a comprehensive introductory treatment, see Anderson and May, 1991.

fever—a mosquito-borne viral disease that is common in the tropics and subtropics around the world); and can include vector population dynamics (e.g., mosquito vectors for dengue, chikungunya, Zika, or malaria; rat vectors for plague or hantavirus).

The inputs used for basic population models include transmissibility,[4] population contact rates,[5] R_0,[6] incubation period, period of infectiousness, duration of illness, and fatality rate. One of these factors (R_0) is especially important in infectious disease modeling, especially for population models. R_0 refers to reproduction rate, or contagiousness, of a disease. It is the expected number of infections caused by a single sick individual in a population otherwise free of the disease (i.e., at the time of disease introduction, when no one has yet been exposed). R_0 can be understood as a product of the disease transmissibility and the population contact rates. For example, rural populations may have a lower R_0 than city dwellers for influenza because population contact rates are lower. A disease with $R_0 > 1$ can spread through the population rapidly (because each infection leads to more than one additional infection, on average) but may affect some areas or ages more than others, while if $R_0 < 1$, the disease will not continue to spread through that population. For example, pinkeye (also called conjunctivitis) may infect an entire preschool but is unlikely to spread to, or at least beyond, the parents or caretakers of those children. This means that the overall R_0 may be close to 1, but the R_0 for young children is much higher. A very contagious disease, such as measles, has a relatively high R_0 (12 to 18), which can lead to extensive spread in an unvaccinated population. Ebola is far less contagious, with an R_0 of approximately 2. Even less-contagious pathogens, such as seasonal influenza most years or even pandemic influenza, have R_0 values just slightly above 1. In contrast to entirely new pandemic influenza strains, seasonal influenza spreads less extensively because many people are immune by virtue of vaccination.

Estimating and using R_0 is also a helpful example of how models use different information. The appropriate R_0 value can be found algebraically given the relevant component data, or it can be estimated in a variety of ways. For instance, statistical methods can be used to find the number of secondary cases based on averages from detailed data, while epidemiological models can estimate it based on population spread rates, such as simulating population models to find the value that best fits the data.[7] Different models use R_0 as an input in two fundamentally different ways: Some use it as a parameter in modeling the disease spread, and others use it as a threshold that determines whether an outbreak will propagate or stall.

The dynamics among the various modeling inputs are represented as a system of (ordinary[8]) differential equations in which each compartment is represented by a number of people

---

[4] Transmissibility is usually expressed in units of rate (e.g., 1/time). R_0, which is unitless, is computed from this rate and the progression rates.

[5] Population contact rates are how often the members in a pair of compartments interact with each other. The population contact rate is expressed in interactions per unit of time for each pair, forming a contact matrix.

[6] R_0 is typically pronounced "R-naught" or "R-zero."

[7] Klaus Dietz, "The Estimation of the Basic Reproduction Number for Infectious Diseases," *Statistical Methods in Medical Research*, Vol. 2, No. 1, 1993.

[8] There is a distinction between so-called "ordinary" differential equations and more-complex partial differential equations, stochastic differential equations, and so on. This topic has a rich literature. See, for example, Robert Smith?, *Modelling Disease Ecology with Mathematics*, American Institute of Mathematical Sciences, Series on Differential Equations and Dynamical Systems, Vol. 2, December 15, 2008; Emilia Vynnycky and Richard G. White, *An Introduction to Infectious Disease Modelling*, Oxford, UK: Oxford University Press, 2010; Avner Friedman and Chiu-Yen Kao, *Mathematical Modeling*

with the indicated status at each point in time. In a simple case, the model shows that, as the number of infected individuals (I) increases, the rate at which new individuals are exposed goes up by a rate determined by transmissibility, and these exposed individuals become infected at a rate determined by the incubation rate. These infected individuals then drive further exposure of susceptible people, leading to disease spread.

Ordinary differential equation (ODE) population models are manageable from a mathematical perspective and computationally fast to run because they use computational techniques that represent large groups of people (compartments) compactly, simplify the disease transitions, and, in their simple form, do not account for uncertainty. Population models with very few compartments allow modelers to obtain an initial approximation of the disease dynamics. However, their computational speed and simplicity can sometimes come at a cost of being less accurate. Single runs of an ODE population model are *deterministic*, meaning that the same inputs always produce the same outputs. ODE models cannot represent or accommodate uncertainty. They implicitly assume homogeneous mixing, meaning that all people are equally likely to interact with any other individual, across geography, groups, and other classification categories. The disease transitions are exponential, which simplifies the disease process in ways that distort the timing of the disease progression, leading to situations where parts of the population transition from susceptible to exposed to infected in near-zero time. Last, ODE models typically use static parameters, which do not include, for example, seasonal components or behavioral change over time; however, such models can take seasons into account—for example, by varying the transmissibility parameter for different time periods.

Modern population models have addressed each of these limitations in different ways,[9] though at a cost, with higher complexity and greater computation requirements, among other factors. For example, to add uncertainty to deterministic results, analysts can use such techniques as sampling model input parameter values within a given estimation range. By running the model many times instead of just once, analysts can use these techniques to show how the results change if the inputs are varied slightly. Sensitivity analyses can then determine the leverage of each input parameter on the model outputs. The most obvious cost of any of these techniques is that they take more time—although they are usually still fast compared with most other modeling approaches. Some more-complex models also sacrifice some of the simplicity of the system by replacing ODEs with (more-complex) *stochastic* (chance-driven, rather than deterministic) differential equations. These equations eliminate the assumption of exponential transition rates and allow for the relationship between compartments to include random effects. In order to relax the simplification of homogeneous mixing, additional compartments can be used to represent the groups with different characteristics, and this can also account for other factors affecting the likelihood of a person contracting or spreading the disease, such as age, profession, and geographical location.

Modeling a more diverse population accurately requires that the population be subdivided into smaller groups of common key characteristics that are relevant to the infection under consideration. This accuracy comes with an additional cost: Models with more compartments to represent detailed population and disease dynamics require more-detailed inputs that

---

of Biological Processes, New York: Springer, 2014; and Morris W. Hirsch, Stephen Smale, and Robert L. Devaney, *Differential Equations, Dynamical Systems, and an Introduction to Chaos*, Waltham, Mass.: Elsevier, 2013.

[9] For more on many of the topics of sensitivity analyses, stochastic differential equations, or forcing, see Smith?, 2008; Vynnycky and White, 2010; Friedman and Kao, 2014; and Hirsch, Smale, and Devaney, 2013.

can be harder to estimate.[10] As a simple example, extending an SIR model to represent multiple population segments (e.g., different age groups or geographic regions) requires estimating the different susceptibility of each group and the contact frequencies between the groups; doubling the number of compartments would square the number of contact frequency estimates needed. Nonstatic parameters are commonly implemented using "forcing" from external variables, so that the (previously static) relationships depend on (are "forced by") externally provided factors (e.g., temperature for a mosquito-borne disease) in order to represent the spread of diseases more accurately.

Population models are the best available tool for understanding the dynamics of disease spread. They can also be used to explore potential effects of untested interventions by comparing predicted model results that assume no intervention ("baseline") with results that introduce the intervention into model inputs. The relative simplicity in building the basic population model and the speed and ability to represent changes in the disease dynamics make these models particularly useful for exploring different features of disease dynamics. On the other hand, as we discussed, the simplifications make them more limited in predictive ability, and they are less useful for modeling heterogeneous population groups.

One example of a population model used for a policy investigation is a RAND model for HIV, used in a prospective evaluation of a test-and-treat policy for testing men who have sex with men in Los Angeles County and treating those who are infected at a much earlier clinical stage.[11] The study explored the likely effects of the policy before it was implemented by using a population model in which the effect of interventions was clearly represented. Because the model was flexible, the model could be run quickly, and the population being studied was mostly homogenous because the scope was limited, relatively compact compartmental modeling allowed exploratory modeling along several potential future paths. Using historical data from 2000 to 2009 and known epidemiological characteristics of HIV to calibrate the model, the study compared future disease spread under the status quo with the future under a test-and-treat policy. It also explored such strategies as increasing testing or treatment individually. This approach was able to estimate the approximate effect of different interventions without waiting for the types of comparison data needed for postimplementation evaluation. In addition, the large-scale, longer-term impacts discussed made exact predictive accuracy—which is difficult to achieve with population models—less critical.

Simulation Models

Simulation models for infectious disease are higher fidelity than population models, but they use a related set of conceptual models.[12] These models expand on the scientific theory basis that underpins population models by explicitly including processes that are only approximated in population models. Simulation models represent individuals or small groups and track their

---

[10] This is a general phenomenon in modeling, frequently called the "curse of dimensionality"; that is, more dimensions and options make the number of variables and cases expand exponentially or combinatorially.

[11] Neeraj Sood, Zachary Wagner, Amber Jaycocks, Emmanuel Drabo, and Raffaele Vardavas, "Test-and-Treat in Los Angeles: A Mathematical Model of the Effects of Test-and-Treat for the Population of Men Who Have Sex with Men in Los Angeles County," *Clinical Infectious Diseases*, Vol. 56, No. 12, June 15, 2013.

[12] The methods for solving or simulating the different types of models can differ greatly, but for our purposes, these are computational, rather than modeling, concerns. For this reason, stochastic population models may be solved via mathematical simulation, but we do not consider them simulation models.

status over time. We discuss the advantages and disadvantages in more detail, but the obvious advantage of simulation models is their added flexibility and precision, albeit at the cost of more complexity when building the models and slower speed when running them.

Simulation models can be further categorized as *microsimulation models* or *agent-based models*. The difference is subtle, but in general, microsimulation models use "top-down" exogenous factors (e.g., contact rates, social structures), while agent-based models generally use "bottom-up" endogenous factors (e.g., how people's contact frequency or behavior responds to the disease outbreak).[13] These models, particularly microsimulation models, are heavily used in health research outside the context of communicable disease—for example, in research on health system capacity,[14] health behaviors,[15] insurance,[16] mental health in the military,[17] and noncommunicable diseases.[18] With our present focus on infectious diseases, we do not attempt to address the full range of ways in which the models described here are used.

Microsimulation models typically represent individuals explicitly and differ from agent-based models in using mainly exogenous rather than endogenous factors. They use empirically derived data about disease spread to more accurately represent the spread of the disease over time, allowing a much richer, dynamic evolution of the disease spread. The empirical data can be derived from statistical regression or from other traditional statistical models using available data.[19]

Microsimulation models for infectious diseases define *events* that individuals may experience, instead of defining a fixed set of possible compartments or states, the way theory-based population, or compartmental, models do. This allows for the representation of a state space that is not limited by compartmental considerations. For example, a microsimulation model may include a variable for how infectious an individual is, instead of assigning the individual to the category of "infectious" (as in a population model) or representing the number of people with whom someone interacts (as in an agent-based model). The value can be a function of how long it has been since the individual was exposed to the disease, as well as other individual characteristics, such as antiviral drug usage, adherence to therapy, age, and social or sexual

---

[13] The terms *exogenous* (literally, generated externally) and *endogenous* (generated internally) are heavily used in economics and the literature about modeling. How the different types of inputs are important in these models is explained more later.

[14] See, for example, Dana P. Goldman, David M. Cutler, Paul G. Shekelle, Jay Bhattacharya, Baoping Shang, Geoffrey F. Joyce, Michael D. Hurd, Dawn Matsui, Sydne Newberry, Constantijn (Stan) Panis, Michael W. Rich, Catherine K. Su, Emmett B. Keeler, Darius N. Lakdawalla, Michael E. Chernew, Feng Pan, Eduardo Ortiz, Robert H. Brook, Alan M Garber, and Shannon Rhodes, *Modeling the Health and Medical Care Spending of the Future Elderly*, Santa Monica, Calif.: RAND Corporation, RB-9324, 2008.

[15] See, for example, Jonathan P. Caulkins, Rosalie Liccardo Pacula, Susan M. Paddock, and James Chiesa, *School-Based Drug Prevention: What Kind of Drug Use Does It Prevent?* Santa Monica, Calif.: RAND Corporation, MR-1459-RWJ, 2002.

[16] See, for example, RAND Corporation, "Comprehensive Assessment of Reform Efforts (COMPARE)," web page, undated b.

[17] See Beau Kilmer, Christine Eibner, Jeanne S. Ringel, and Rosalie Liccardo Pacula, "Invisible Wounds, Visible Savings? Using Microsimulation to Estimate the Costs and Savings Associated with Providing Evidence-Based Treatment for PTSD and Depression to Veterans of Operation Enduring Freedom and Operation Iraqi Freedom," *Psychological Trauma: Theory, Research, Practice, and Policy*, Vol. 13, No. 2, June 2011.

[18] See, for example, Sarah Nowak and Andrew M. Parker, "Social Network Effects of Nonlifesaving Early-Stage Breast Cancer Detection on Mammography Rates," *American Journal of Public Health*, Vol. 104, No. 12, December 2014, pp. 2439–2444.

[19] For a more complete overview of these models in a variety of areas, see O'Donoghue, 2014.

behavior; such rates can be derived from medical research or clinical data. This allows for individuals to have a range of characteristics without explicitly modeling compartments for each combination, although it does require much more computational work to represent individuals (in a microsimulation model) than aggregated populations (in a population model).

In order to represent these characteristics, a large number of (possibly correlated) variables are needed, and these different variables can be difficult to collect, making these models difficult to parameterize—although the simplistic assumptions used in population models are hiding, not solving, this problem. These models are considered more predictive, but even in the best case, they require simplifications because of limited knowledge about the population and disease.

Because of the richness of the representation, microsimulation models are relatively computationally complex and slow. They are also expensive to build because there are so many individuals or agents included, and each one requires data collection or programming. Such models also do not always scale up in size easily, because the number of individuals increases linearly, but the computational work frequently increases nonlinearly. Thus, this type of model is frequently run on supercomputers when representing larger scales, especially for modeling entire countries or systems that have complex interactions between individuals. These factors can make exploratory work and uncertainty quantification more difficult.

As an example of the microsimulation modeling approach for infectious disease, researchers considered different scenarios reflecting a potential bioterrorist attack using smallpox virus, including patient isolation and different options related to targeting and timing of the smallpox vaccination (all aimed at preventing further spread of the virus).[20] The study used an event-driven simulation to track individuals in each disease generation. The microsimulation model considered prior vaccination of health workers and the general public and assessed the degree of disease spread and number of expected deaths under each smallpox scenario. The model represented the early spread, which enabled a small scope and therefore reduced computational needs. This initial microsimulation model was then used to build and validate a much simpler population model, which enabled more-rapid exploration of the policy options available. The modeling analyses favored prior vaccination of health workers under most scenarios, but the analyses favored prior vaccination of the general public only if the likelihood of an attack or multiple attacks was high. These analyses were instrumental in U.S. disaster preparedness planning in terms of both vaccine stockpiling and likely responses in the event of a smallpox emergency.

Agent-based simulation models, also sometimes called individual-based models, represent the behavior of individuals and their disease status as it results from this behavior. Early work on agent-based models occurred outside the realm of disease modeling, specifically in economics and ecology, but this modeling approach has proven fruitful in a wide variety of contexts.[21] Agent-based models can be considered a special case of microsimulation models, but the typi-

---

[20] S. A. Bozzette, R. Boer, V. Bhatnagar, J. L. Brower, E. B. Keeler, S. C. Morton, and M. A. Stoto, "A Model for Smallpox-Vaccination Policy," *New England Journal of Medicine*, Vol. 348, 2003.

[21] For a more complete introduction to agent-based modeling in general, see Joshua M. Epstein, *Generative Social Science: Studies in Agent-Based Computational Modeling*, Princeton, N.J.: Princeton University Press, 2007; the book includes chapters on disease modeling and on the limitations and advantages of agent-based models in this and other contexts. See also S. F. Railsback and V. Grimm, *Agent-Based and Individual-Based Modeling: A Practical Introduction*, Princeton, N.J.: Princeton University Press, 2012; and A. M. El-Sayed, P. Scarborough, L. Seemann, and S. Galea, "Social Network Analysis and Agent-Based Modeling in Social Epidemiology," *Epidemiologic Perspectives and Innovations*, Vol. 9, 2012.

cal focus is different. In these models, instead of using the empirically derived data about disease spread used by microsimulation models, individuals change their behavior (referred to earlier as bottom-up or endogenous factors). This happens according to predefined rules followed by the individuals, and the behaviors can be functions of different endogenous factors. For example, people may interact with other people less when they observe that many other people are infected, slowing the spread of the disease. Similarly, they may be more likely to seek immunization if they see that many people were infected the previous year. This can create "emergent" properties of a system—those that are caused by the combination of endogenous factors—and these properties evolve over time. This model capability allows complex disease dynamics to result from relatively simple rules. In the examples mentioned, changing individual behavior can lead to complex patterns of disease spread, and the resulting simulation can be analyzed to track the spread of disease that emerged from individuals' actions. As a consequence, agent-based models can show when rapidly spreading diseases may be less damaging as a result of behavioral changes that blunt the spread of the disease, or they may exhibit complex multiyear patterns.

Generally, if the system modeled is not already well understood, agent-based models may help scientists explore how disease can spread and what behaviors could give rise to observed features of an epidemic. Therefore, a key component of an agent-based model is a description of the behaviors of individuals and how behaviors may evolve and diffuse in the population as a result of individuals' (or agents') interactions. If the systems are already well understood, this class of model can be a powerful policy tool for understanding behavior in a complex system that is otherwise hard to explore.[22] Examples of the processes that the behavioral module would describe include social distancing, sexual behaviors, propensity to seek vaccination, and treatment.

Because of the strategy of representing sources (not results) of behaviors, agent-based models are sometimes referred to as "naïve" or "direct," in the sense that fundamental characteristics of the population's behavior, such as their medical behaviors or the people with whom they interact, serve as model inputs. For example, if we model the locations and interactions of individuals based on their social networks and geography, an agent-based model can simulate disease spread based on the length or frequency of interactions with infectious individuals. The infectivity of an individual can fluctuate over time based on that individual's behaviors, which can change based on his or her own disease status, social network, and perception of other factors.[23] Changes in individual behavior give rise to changes over the course of the epidemic, without (directly) using observed changes in interaction frequency. Based on the simulation, such emergent properties as R_0 (which are inputs to other types of models) can be calculated from the simulated history of a disease. This allows complex dynamic changes to factors that population models cannot easily represent as dynamically changing.

Generally, the approach taken by agent-based models is to formulate rules describing how behaviors change based on feedback at the population level (macroscopic) and at smaller levels (microscopic). An example of macroscopic feedback is the perceived prevalence of a disease by the individuals in the model who could, as a consequence, change their behaviors. For

---

[22] For example, it may be hard to observe the inputs needed to inform a microsimulation model, but the human behavior that leads to those hard-to-observe inputs may be understood.

[23] The individual's behaviors change endogenously, but the changes can be due to either exogenous factors (e.g., decisions made by policymakers) or endogenous factors (e.g., overall disease prevalence).

example, they could increasingly engage in social distancing. An example of microscopic feedback is the perceived number of individuals who have been vaccinated in their social networks. In contrast to evidence-based microsimulation models, in which a limited set of individuals' behaviors are included, agent-based models consider a more complete set of behaviors even if there are very limited data that can help precisely describe these processes. Instead, where data are limited, behavioral rules are formulated based on heuristics and informed guesses. Therefore, agent-based models formulate both the disease contagion dynamics and the individuals' behaviors based on mechanistic rules rather than statistical regression models. The rules describing the contagion dynamics are based on known biological mechanisms. The rules describing the behavioral dynamics can often be based on postulated mechanisms or heuristics of how agents interact in the real world. For this reason, compared with evidence-based microsimulation models, agent-based models are not very reliable at precisely forecasting dynamics under status quo conditions or under policies that make minor changes to the status quo. However, agent-based models are ideal to explore "what-if" scenarios whereby the policy is based on a behavioral response, including behavioral diffusion and policies that are different from the status quo. For example, incentive-based policies can be tested with an agent-based model precisely because the model encapsulates a behavioral model of the agents.

In addition to many of the same computational considerations mentioned for microsimulation models, key methodological challenges for agent-based models are validity (ensuring that conclusions from the model are correct or reasonable) and verification (ensuring that conclusions are accurate). Because these models are created to represent the process for the overall dynamics of the disease, not to replicate the observed behavior, the outputs of an agent-based model must be tested carefully to ensure that the model functions appropriately, especially outside of the range for which it was calibrated. This may mean that results should be compared with empirical data, if possible, or at least be assessed by experts to ensure that they are reasonable. Another challenge is the correct specification of the ways in which individuals or agents interact. Missing features or incorrect simplifications can be critical in unanticipated ways, so that slight changes in behavior or the addition of new types of individuals or agents can completely change the dynamics of the system.

One example of policy application of an agent-based model is an exploratory project led by one of the authors of this report (Raffaele Vardavas) on behaviors toward vaccination based on different features of individuals' social networks.[24] The project focused on understanding the interplay between the social network structure, agents' yearly vaccination behaviors, and influenza epidemiology. The project modeled the coupled dynamics of influenza transmission and vaccination behavior to test the effects of different behavioral interventions. This required explicit modeling of social networks, which is ideally suited to agent-based simulation modeling. By limiting the scale of the model to specific social networks—in which, for simplicity, the network links connecting individuals were assumed to be independent from socioeconomic, location, and demographic attributes—the model was able to explore these dynamics. This type of model does not claim to be predictive or accurate at a detailed level. However, it was able to reproduce stylized facts of influenza epidemiology, including observed U.S. levels of yearly vaccination coverage, the yearly cumulative incidence, and observed distributions regarding the propensity to vaccinate. Results from the model provided initial insights

---

[24] Vardavas and Marcum, 2013.

into the likelihood of success of incentive-based interventions to vaccinate that rely on social network influences. This effort used a combination of models that facilitated a better understanding of relevant parameters; it used a social network of contact between agents for disease spread instead of making simplified assumptions about contact between large-scale population groups (the latter typifying population models, as described earlier). The contact network was also used to model how people make choices about vaccination behavior. Those who are closely connected by their social networks to others who get vaccinated are more likely to get vaccinated. Collective decisions about vaccination drive influenza epidemiology that, in turn, affects future individual-level decisions. To support this work, many behavioral model parameters that are needed to characterize agent behavior—which are difficult to find in extant literature—were quantified using results from a survey on influenza behavior run through the RAND American Life Panel.[25]

Statistical Models

Statistical models are distinct from theory-based models, although, in practice, they may overlap or be used in combination with them. Statistical models are typically distinguished by use of mathematical relationships to directly represent quantities of interest, relying on mining large amounts of data. Statistical models can reproduce dynamic real-world relationships by learning trends from empirical data and encoding these dependencies in a mathematical model, without directly representing the causal scientific factors involved.[26] These models use real-world outputs (i.e., the empirical "training" data set[27]), which allows the statistical method to learn the correct behavior (e.g., disease occurrence).

The learned mathematical relationships, or patterns, can forecast disease occurrence. Statistical models of all types enable this data-driven, inductive prediction under limited model assumptions, despite potentially limited exact theoretical understanding. Because of this, practically all inductive learning includes an implicit assumption of *invariance*,[28] meaning that the past relationship between the disease characteristics and occurrence continues to tell us about future outbreaks.

We can take this further in the context of decision support: Models should be used to inform decisions only in scenarios similar to the context in which the model was built. To the extent that the scenarios differ, any inference is less valid. For example, if a model uses data from only one location or with just one type or intensity of intervention, using that model to decide a more general question can be problematic.

---

[25] The RAND American Life Panel is a nationally representative, probability-based panel of more than 6,000 adults who are regularly interviewed for research purposes. For more information, see RAND Corporation, "American Life Panel," web page, undated a.

[26] Statistical models do not naturally distinguish events that are causally connected from events that are merely correlated. This is because the observational data (recording observed events) can account only for historically observed scenarios, *not* for scenarios that did not occur, such as what would have happened if a new intervention had been applied. Because of this, causality in statistical models requires econometric approaches to inferring causality from quasi-experimental data, and a large literature exists on how this is done.

[27] The use of example data to train models is referred to as *supervised learning* (especially in machine-learning literature). It is standard practice for many types of statistical models used for disease modeling.

[28] See Valiant, 2013.

Statistical models are especially susceptible to this kind of problem because of their strong dependence on the training data. Additionally, there is a related, but more fundamental, problem specific to typical applications of statistical models. This is the problem of inferring causality: How can we infer a causal link between two events using statistical methods? This is often critical for questions of the relative merits of interventions (alternative pasts or future forecasts). For example, would dengue incidence drop if citizens were fined for public health violations (e.g., harboring mosquito breeding sites in their yards during the rainy season)? If a public health media campaign had not been run, would vaccination rates for a vaccine-preventable disease, such as influenza, be different from what was observed? Statistical models have difficulty answering these questions.

We describe two general classes of statistical models—traditional *regression-based models* and modern *machine-learning models*.

Regression-Based Models

Regression-based models have been extensively used to model the real-world courses of diseases and consequently can be considered as a standard or traditional statistical approach. An example of these is the common *least-squares regression model*, which Carl Friedrich Gauss introduced in the late 1700s while trying to predict a comet's path.[29] These models propose a specific mathematical formula that depends on a set of parameters for the relationship between the output variable and a set of potential input variables, or predictors. This formula accounts for noise or corruption in the data using statistical methods of analysis. Supervised learning methods select relevant predictors and produce an optimal fit for the proposed parametric form. Modeling skill and empirical or statistical constraints inform the choice of parametric forms and potential predictors.

These learned mathematical relationships provide useful indications of the future course of the disease even if there is no underlying theory. They are predictions for the future, not explanations of the disease, so they may not tell the full story. For example, it is standard to use such regression-based models to build experimental support for new scientific hypotheses.[30] However, given the dependence on the invariance assumption, the relationships in these models may not have full explanatory power without established context-relevant scientific theory. Traditional regression-based methods can also be limited in the range of relationships they can model. Despite these limitations, such models are computationally very fast and can be used to model and understand the direction and magnitude of relationships in the data even when the scientific relationships are unknown. This means that quick and roughly accurate models can predict outcomes even when the causes are not known or understood or when the scientific theory is insufficient for representation.[31]

---

[29] In least-squares regression, the "best" values are those that minimize the sum of the square of the differences between the actual value and the value output by the formula.

[30] For example, the determination of half-lives and decay profiles for radioisotopes often involves a least-squares fit of observed decay data to statistical Poisson emission models. Pharmacokinetic analyses also use statistical models to model the bioavailability of drugs.

[31] Early in the recent Ebola epidemic, the social factors leading to the spread of Ebola were only partially known (for example, the role of unsafe funeral practices and sexual contact), but predictive statistical models can be effective despite this gap in scientific knowledge.

Examples of this approach include using the Poisson regression[32] and related regressions to predict mosquito populations and dengue fever incidence in tropical locations.[33] These predictions are based on population-level models that mine data sets for statistical relationships between disease-relevant variables (e.g., environmental conditions, population factors, policy intervention) and disease incidence.

Machine-Learning Models

Machine-learning models examine observed input and output data and use algorithmic statistical techniques to "learn" structured relationships in the data. Modern machine-learning methods extend the expressive range of regression-based statistical methods to include more-complex dynamics. They can use more-complex adaptive mathematical structures to represent arbitrarily complex relationships.[34] Therefore, their usefulness for disease forecasting and decision support is unsurprising. The basic approach is to use sample input-output data to learn complex relationships, often with the goal of predicting future outputs. The complexity of the relationships makes them more predictive but often harder to understand or explain. There are several powerful, versatile, and robust families of machine-learning models. We recently demonstrated the use of one such model—an *artificial neural network (ANN) model*—for forecasting the incidence of dengue fever. Here, we use this as a single example of this class of model while noting the benefits and drawbacks of machine-learning methods more generally.

ANN models owe their inception to attempts in the past 80 years to understand and mimic information-processing in biological systems. Networks of biological neurons (e.g., the mammalian brain) have evolved the ability to learn patterns from examples in noisy, uncertain environments.[35] The artificial neurons (or nodes) and their interconnections of ANNs have vastly simplified dynamics compared with their biological counterparts (neurons and synapses, respectively). But ANN models have proven to be a versatile tool for learning statistical patterns in many domains.[36] For example, ANNs have found supervised learning applications in computer vision, speech recognition, economic prediction, and automated control, among others.

---

[32] The proposed parametric form for a Poisson regression is a Poisson distribution, in which the conditional mean is a polynomial function of the predictors. In simplified terms, it assumes that there is some linear relationship between the logarithm of the number of cases and the values of different environmental variables, and the method finds the relationship of available environmental variables that best predicts the local rate of dengue incidence.

[33] See, for example, S. Wongkoon, M. Jaroensutasinee, and K. Jaroensutasinee, "Distribution, Seasonal Variation & Dengue Transmission Prediction in Sisaket, Thailand," *Indian Journal of Medical Research*, Vol. 138, No. 3, 2013; and B. M. Althouse, Y. Y. Ng, and D. A. Cummings, "Prediction of Dengue Incidence Using Search Query Surveillance," *PLoS Neglected Tropical Diseases*, Vol. 5, No. 8, 2011.

[34] Model families that can represent relationships are termed *universal approximators* if they can learn any (arbitrarily complex) statistical input-output relationship when given enough examples.

[35] The ANN is a crude approximation of its biological counterpart. The mechanics of common ANN training algorithms may not be similar to the biological mechanics of learning in the mammalian brain. See Bart Kosko, *Neural Networks and Fuzzy Systems: A Dynamical Systems Approach to Machine Intelligence*, Vol. 1, Englewood Cliffs, N.J.: Prentice Hall, 1992.

[36] ANN models are universal approximators, as defined earlier. See H. White, "Learning in Artificial Neural Networks: A Statistical Perspective," *Neural Computation*, Vol. 1, No. 4, 1989; and K. Hornik, M. Stinchcombe, and H. White, "Multilayer Feedforward Networks Are Universal Approximators," *Neural Networks*, Vol. 2, No. 5, 1989.

We recently demonstrated the use of an ANN modeling approach for forecasting the incidence of dengue fever based on historical training data from two cities—San Juan, Puerto Rico, and Iquitos, Peru. This exercise was in response to a joint epidemic-forecasting challenge from federal agencies, including the Centers for Disease Control and Prevention, the U.S. Department of Defense, and the National Oceanic and Atmospheric Administration.[37] Prior literature review and some exploratory data analysis indicated that temperature, rainfall, and recent incidence history were considered the primary factors predictive of future dengue incidence.[38]

Back-propagation methods tune the ANN model to learn predictive statistical relationships from past input-output examples.[39] In our exploration of dengue in San Juan and Iquitos, we configured our ANN models as a time-series predictor to forecast future dengue incidence using a short-term history of recent incidence, temperature, and rainfall as inputs.[40] We divided the environmental and (human case) incidence data into a collection of time windows. The ANN uses these historical examples to learn the relationship between input time windows (consisting of recent history of incidence, temperature, and rainfall) and the output (future dengue incidence over a window of two to six weeks). The model can also provide limited what-if forecasts if we input hypothetical values for recent incidence and environmental variables. This means that we can simulate alternative futures by presenting the model with hypothetical input values. The main caveat here is that the model outputs will be unreliable if the hypothetical input values fall far outside the typical input range seen during training. The results of any type of model should be subject to skepticism when dealing with atypical inputs.

The precise functional dependence in our dengue model was not easy to tease apart. This is a common feature in machine-learning models. As noted earlier, the model uses recent data on dengue incidence, local temperature, and local precipitation to forecast future dengue incidence. Such factors as mosquito spraying in some or all of one or both of these cities were not included and thus could easily mask more-fundamental underlying disease patterns. The historical input variables are lagged and truncated time-series vectors for the different predictors. Back-propagation learning tunes the connections between the nodes to reproduce the relationship between the input variables (temperature, precipitation, incidence) and the output forecasts (future incidence) based on past examples in the data set.

---

[37] Centers for Disease Control and Prevention, "Combating Dengue with Infectious Disease Forecasting," press release, June 5, 2015.

[38] See, for example, L. M. Rueda, K. J. Patel, R. C. Axtell, and R. E. Stinner, "Temperature-Dependent Development and Survival Rates of Culex Quinquefasciatus and Aedes Aegypti (Diptera: Culicidae)," *Journal of Medical Entomology*, Vol. 27, No. 5, 1990; and J. B. Siqueira, Jr., C. M. T. Martelli, G. E. Coelho, A. C. da Rocha Simplício, and D. L. Hatch, "Dengue and Dengue Hemorrhagic Fever, Brazil, 1981–2002," *Emerging Infectious Diseases*, Vol. 11, No. 1, 2005.

[39] Back-propagation is the standard supervised learning method for ANNs. In general, supervised learning algorithms for machine-learning models are similar in approach to those used in traditional statistical models (e.g., least-squares regression): They attempt to find a "best fit" for the model parameters based on errors between the model output and the actual value observed.

[40] The model structure is a hybrid of time-series prediction and sensor fusion. The model's time-series predictor structure generalizes simple time-series models common in econometrics. And the model's use of different types of sensory inputs is reminiscent of *information fusion* methods, such as Kalman filters used in target-tracking and motion control for robots.

References

Althouse, B. M., Y. Y. Ng, and D. A. Cummings, "Prediction of Dengue Incidence Using Search Query Surveillance," *PLoS Neglected Tropical Diseases*, Vol. 5, No. 8, 2011, p. e1258.

Anderson, R. M., and R. M. May, *Infectious Diseases of Humans: Dynamics and Control*, Oxford, UK: Oxford University Press, 1991.

Bardach, Eugene, *A Practical Guide for Policy Analysis: The Eightfold Path to More Effective Problem Solving*, 4th ed., Thousand Oaks, Calif.: CQ Press, 2012.

Bottou, L., J. Peters, J. Quinonero-Candela, D. X. Charles, D. M. Chickering, E. Portugaly, D. Ray, P. Simard, and E. Snelson, "Counterfactual Reasoning and Learning Systems: The Example of Computational Advertising," *Journal of Machine Learning Research*, Vol. 14, No. 1, 2013, pp. 3207–3260.

Bozzette, S. A., R. Boer, V. Bhatnagar, J. L. Brower, E. B. Keeler, S. C. Morton, and M. A. Stoto, "A Model for Smallpox-Vaccination Policy," *New England Journal of Medicine*, Vol. 348, 2003, pp. 416–425.

Brugnach, M., C. Pahl-Wostl, K. E. Lindenschmidt, J. A. E. B. Janssen, T. Filatova, A. Mouton, G. Holtz, P. van der Keur, and N. Gaber, *Complexity and Uncertainty: Rethinking the Modelling Activity*, Lincoln, Neb.: U.S. Environmental Protection Agency Paper 72, 2008. As of April 25, 2016: http://digitalcommons.unl.edu/cgi/viewcontent.cgi?article=1071&context=usepapapers

Califano, Joseph, Jr., "Introduction," in Richard E. Neustadt and Harvey V. Fineberg, *The Swine Flu Affair: Decision-Making on a Slippery Disease*, U.S. Department of Health, Education, and Welfare, 1978.

Caulkins, Jonathan P., Rosalie Liccardo Pacula, Susan M. Paddock, and James Chiesa, *School-Based Drug Prevention: What Kind of Drug Use Does It Prevent?* Santa Monica, Calif.: RAND Corporation, MR-1459-RWJ, 2002. As of April 25, 2016: http://www.rand.org/pubs/monograph_reports/MR1459.html

Centers for Disease Control and Prevention, "Combating Dengue with Infectious Disease Forecasting," press release, June 5, 2015. As of May 10, 2016: http://www.noaanews.noaa.gov/stories2015/20150605_Dengue-Forecast-PR1.pdf

Chamberlin, Margaret, Shira Efron, and Melinda Moore, *A Simple Approach to Assessing Potential Health Emergency Interventions: A Proof of Concept and Illustrative Application to the 2014–2015 Ebola Crisis*, Santa Monica, Calif.: RAND Corporation, PE-148-RC, 2015. As of April 25, 2016: http://www.rand.org/pubs/perspectives/PE148.html

Chamberlin, Margaret, Adeyemi Okunogbe, Melinda Moore, and Mahshid Abir, *Intra-Action Report—A Dynamic Tool for Emergency Managers and Policymakers: A Proof of Concept and Illustrative Application to the 2014–2015 Ebola Crisis*, Santa Monica, Calif.: RAND Corporation, PE-147-RC, 2015. As of April 25, 2016: http://www.rand.org/pubs/perspectives/PE147.html

Choisy, M., P. Sitboulang, M. Vongpanhya, C. Saiyavong, B. Khamphaphongphanh, B. Phommasack, F. Quet, Y. Buisson, J.-D. Zucker, and W. van Pahuis, "Rescuing Public Health," in Serge Morand, Jean-Pierre Dujardin, Régine Lefait-Robin, and Chamnarn Apiwathnasorn, eds., *Socio-Ecological Dimensions of Infectious Diseases in Southeast Asia,* Springer-Verlag, 2015, pp. 171–190.

Dalal, S., D. Khodyakov, R. Srinivasan, S. Straus, and J. Adams, "ExpertLens: A System for Eliciting Opinions from a Large Pool of Non-Collocated Experts with Diverse Knowledge," *Technological Forecasting and Social Change*, Vol. 78, No. 8, 2011, pp. 1426–1444.

Dausey, D. J., J. W. Buehler, and N. Lurie, "Designing and Conducting Tabletop Exercises to Assess Public Health Preparedness for Manmade and Naturally Occurring Biological Threats," *BMC Public Health*, Vol. 7, 2007, p. 92.

Dausey D. J., and M. Moore, "Using Exercises to Improve Public Health Preparedness in Asia, the Middle East and Africa," *BMC Research Notes*, Vol. 7, 2014, p. 474. As of April 25, 2016:
http://www.biomedcentral.com/1756-0500/7/474

Dietz, Klaus, "The Estimation of the Basic Reproduction Number for Infectious Diseases," *Statistical Methods in Medical Research*, Vol. 2, No. 1, 1993, pp. 23–41. As of May 10, 2016:
http://smm.sagepub.com/content/2/1/23.full.pdf

Epstein, Joshua M., *Generative Social Science: Studies in Agent-Based Computational Modeling*, Princeton, N.J.: Princeton University Press, 2007.

Frieden, Tom, "Staying Ahead of the Curve: Modeling and Public Health Decision-Making," presentation, CDC Grand Rounds, January 19, 2016.

Friedman, Avner, and Chiu-Yen Kao, *Mathematical Modeling of Biological Processes*, New York: Springer, 2014.

Goldman, Dana P., David M. Cutler, Paul G. Shekelle, Jay Bhattacharya, Baoping Shang, Geoffrey F. Joyce, Michael D. Hurd, Dawn Matsui, Sydne Newberry, Constantijn (Stan) Panis, Michael W. Rich, Catherine K. Su, Emmett B. Keeler, Darius N. Lakdawalla, Michael E. Chernew, Feng Pan, Eduardo Ortiz, Robert H. Brook, Alan M. Garber, and Shannon Rhodes, *Modeling the Health and Medical Care Spending of the Future Elderly*, Santa Monica, Calif.: RAND Corporation, RB-9324, 2008. As of April 25, 2016:
http://www.rand.org/pubs/research_briefs/RB9324.html

Heckman, J. J., "Econometric Causality," *International Statistical Review*, Vol. 76, No. 1, 2008, pp. 1–27.

Hirsch, Morris W., Stephen Smale, and Robert L. Devaney, *Differential Equations, Dynamical Systems, and an Introduction to Chaos*, Waltham, Mass.: Elsevier, 2013.

Hornik, K., M. Stinchcombe, and H. White, "Multilayer Feedforward Networks Are Universal Approximators," *Neural Networks*, Vol. 2, No. 5, 1989, pp. 359–366.

Kilmer, Beau, Christine Eibner, Jeanne S. Ringel, and Rosalie Liccardo Pacula, "Invisible Wounds, Visible Savings? Using Microsimulation to Estimate the Costs and Savings Associated with Providing Evidence-Based Treatment for PTSD and Depression to Veterans of Operation Enduring Freedom and Operation Iraqi Freedom," *Psychological Trauma: Theory, Research, Practice, and Policy*, Vol. 13, No. 2, June 2011, pp. 201–211. As of May 10, 2016:
http://www.rand.org/pubs/external_publications/EP201100107.html

Kosko, Bart, *Neural Networks and Fuzzy Systems: A Dynamical Systems Approach to Machine Intelligence*, Vol. 1, Englewood Cliffs, N.J.: Prentice Hall, 1992.

Lee, B. Y., L. A. Haidari, and M. S. Lee, "Modelling During an Emergency: The 2009 H1N1 Influenza Pandemic," *Clinical Microbiology and Infection*, Vol. 19, No. 11, November 2013, pp. 1014–1022.

Lewis, D., "Causation," *Journal of Philosophy*, 1973, pp. 556–567.

Lipsitch, M., L. Finelli, R. T. Heffernan, G. M. Leung, and S. C. Redd, "Improving the Evidence Base for Decision Making During a Pandemic: The Example of 2009 Influenza A/H1N1," *Biosecurity and Bioterrorism: Biodefense Strategy, Practice, and Science*, Vol. 9, No. 2, June 2011, pp. 89–115. As of April 25, 2016:
http://www.ncbi.nlm.nih.gov/pmc/articles/PMC3102310/

Lurie, N., D. J. Dausey, T. Knighton, M. Moore, S. Zakowsky, and L. Deyton, "Community Planning for Pandemic Influenza: Lessons from the VA Health Care System," *Disaster Medicine and Public Health Preparedness*, Vol. 2, No. 4, 2008, pp. 251–257.

Meltzer, Martin, "What Do Policy Makers Expect from Modelers During a Response?" presentation, CDC Grand Rounds, January 19, 2016.

Meyers, Lauren Ancel, "Modeling to Support Outbreak Preparedness, Surveillance and Response," presentation, CDC Grand Rounds, January 19, 2016.

Moore, Melinda, Edward Chan, Nicole Lurie, Agnes Gereben Schaefer, Danielle M. Varda, and John A. Zambrano, "Strategies to Improve Global Influenza Surveillance: A Decision Tool for Policymakers," *BMC Public Health*, Vol. 8, 2008, p. 186.

Morgan, M. Granger, and Max Henrion, *Uncertainty: A Guide to Dealing with Uncertainty in Quantitative Risk and Policy Analysis*, Cambridge, UK: Cambridge University Press, 1990.

Neustadt, Richard E., and Harvey V. Fineberg, *The Swine Flu Affair: Decision-Making on a Slippery Disease*, U.S. Department of Health, Education, and Welfare, 1978.

Nowak, Sarah, and Andrew M. Parker, "Social Network Effects of Nonlifesaving Early-Stage Breast Cancer Detection on Mammography Rates," *American Journal of Public Health*, Vol. 104, No. 12, December 2014, pp. 2439–2444. As of April 25, 2016:
http://ajph.aphapublications.org/doi/pdf/10.2105/AJPH.2014.302153

O'Donoghue, Cathal, *Handbook of Microsimulation Modelling*, Bingley, UK: Emerald Group Publishing, 2014.

Pascale, Richard T., Jerry Sternin, and Monique Sternin, *The Power of Positive Deviance: How Unlikely Innovators Solve the World's Toughest Problems*, Boston: Harvard Business Press, 2010.

Pearl, J., *Causality*, Cambridge, UK: Cambridge University Press, 2009.

Quade, E.S., *Analysis for Public Decisions*, New York: Elsevier, 1975.

Railsback, S. F., and V. Grimm, *Agent-Based and Individual-Based Modeling: A Practical Introduction*, Princeton, N.J.: Princeton University Press, 2012.

RAND Corporation, "American Life Panel," web page, undated a. As of April 25, 2016:
https://alpdata.rand.org

———, "Comprehensive Assessment of Reform Efforts (COMPARE)," web page, undated b. As of May 10, 2016:
http://www.rand.org/health/projects/compare.html

———, "Delphi Method," web page, undated c. As of April 25, 2016:
http://www.rand.org/topics/delphi-method.html

Rosenbaum, P. R., and D. B. Rubin, "The Central Role of the Propensity Score in Observational Studies for Causal Effects," *Biometrika*, Vol. 70, No. 1, 1983, pp. 41–55.

Rueda, L. M., K. J. Patel, R. C. Axtell, and R. E. Stinner, "Temperature-Dependent Development and Survival Rates of Culex Quinquefasciatus and Aedes Aegypti (Diptera: Culicidae)," *Journal of Medical Entomology*, Vol. 27, No. 5, 1990, pp. 892–898.

Savoia, E., M. A. Testa, P. D. Biddinger, R. O. Cadigan, H. Koh, P. Campbell, and M. A. Stoto, "Assessing Public Health Capabilities During Emergency Preparedness Tabletop Exercises: Reliability and Validity of a Measurement Tool," *Public Health Reports*, Vol. 124, No. 1, 2009, pp. 138–148.

El-Sayed, A. M., P. Scarborough, L. Seemann, and S. Galea, "Social Network Analysis and Agent-Based Modeling in Social Epidemiology," *Epidemiologic Perspectives and Innovations*, Vol. 9, 2012.

Siqueira, J. B., Jr., C. M. T. Martelli, G. E. Coelho, A. C. da Rocha Simplício, and D. L. Hatch, "Dengue and Dengue Hemorrhagic Fever, Brazil, 1981–2002," *Emerging Infectious Diseases*, Vol. 11, No. 1, 2005, p. 48.

Smith?, Robert, *Modelling Disease Ecology with Mathematics*, American Institute of Mathematical Sciences, Series on Differential Equations and Dynamical Systems, Vol. 2, December 15, 2008.

Sood, Neeraj, Zachary Wagner, Amber Jaycocks, Emmanuel Drabo, and Raffaele Vardavas, "Test-and-Treat in Los Angeles: A Mathematical Model of the Effects of Test-and-Treat for the Population of Men Who Have Sex with Men in Los Angeles County," *Clinical Infectious Diseases*, Vol. 56, No. 12, June 15, 2013, pp. 1789–1796.

Sutherland, William J., and Mark Burgman, "Policy Advice: Use Experts Wisely," *Nature*, Vol. 536, October 2015, pp. 317–318. As of April 25, 2016:
http://www.nature.com/news/policy-advice-use-experts-wisely-1.18539

Thacker, S., "Historical Development," in Steven Teutsch and R. Elliot Churchill, eds., *Principles and Practice of Public Health Surveillance*, New York: Oxford University Press, 2000, pp. 1–16.

Valiant, Leslie, *Probably Approximately Correct: Nature's Algorithms for Learning and Prospering in a Complex World*, New York: Basic Books, 2013.

Van Kerkhove, Maria D., and Neil M. Ferguson, "Epidemic and Intervention Modelling—A Scientific Rationale for Policy Decisions? Lessons from the 2009 Influenza Pandemic," *Bulletin of the World Health Organization*, Vol. 90, No. 4, 2012, pp. 306–310. As of April 25, 2016: http://www.ncbi.nlm.nih.gov/pmc/articles/PMC3324871/

Vardavas, Raffaele, and Christopher Steven Marcum, "Modeling Influenza Vaccination Behaviour Via Inductive Reasoning Games," in Piero Manfredi and Alberto D'Onofrio, eds., *Modeling the Interplay Between Human Behavior and the Spread of Infectious Disease*, New York: Springer-Verlag, 2013.

Vynnycky, Emilia, and Richard G. White, *An Introduction to Infectious Disease Modelling*, Oxford, UK: Oxford University Press, 2010.

Watanabe, Ken, *Problem Solving 101: A Simple Book for Smart People*, New York: Penguin Group, 2009.

White, H., "Learning in Artificial Neural Networks: A Statistical Perspective," *Neural Computation*, Vol. 1, No. 4, 1989, pp. 425–464.

Wongkoon, S., M. Jaroensutasinee, and K. Jaroensutasinee, "Distribution, Seasonal Variation & Dengue Transmission Prediction in Sisaket, Thailand," *Indian Journal of Medical Research*, Vol. 138, No. 3, 2013, p. 347.